Climate Change Remedies

Climate Change Remedies

Injunctive Relief and Criminal Law Responses

Jaap Spier and Ulrich Magnus (Eds.)

eleven
international publishing

Published, sold and distributed by Eleven International Publishing
P.O. Box 85576
2508 CG The Hague
The Netherlands
Tel.: +31 70 33 070 33
Fax: +31 70 33 070 30
e-mail: sales@budh.nl
www.elevenpub.com

Sold and distributed in USA and Canada
International Specialized Book Services
920 NE 58th Avenue, Suite 300
Portland, OR 97213-3786, USA
Tel: 1-800-944-6190 (toll-free)
Fax: +1-503-280-8832
orders@isbs.com
www.isbs.com

Eleven International Publishing is an imprint of Boom uitgevers Den Haag.

ISBN 978-94-6236-080-8
ISBN 978-94-6094-829-9 (E-book)

Printed in The Netherlands

Table of Contents

Preface

Climate change poses significant threats to humanity, other living species and the environment. The challenge to come to grips with these threats accelerates at a great pace. Alarming warnings catch the attention of ever more politicians and senior business people. Thanks to the impressive work done by major international organizations, such as various UN bodies, the UN High Commissioner for Human Rights, the EU, the African Union, the World Bank, IPCC, a large number of NGOs and leading newspapers and magazines, public awareness is growing. But so far, the willingness to translate the awareness into concrete, let alone adequate, action has fallen short. Short-term interests and views, the level playing field, greed and the realities of international politics seem to be an insurmountable obstacle to achieving the bitterly needed global action.

It is unrealistic to assume that legal strategies alone could stem the tide to the full extent needed. But they could contribute to a global solution. They could serve as a welcome vehicle for politicians and business leaders of good will to explain to their voters and shareholders that they must change course. If effective, legal strategies could bring the unwilling to their senses.

The traditional legal paradigm is to find solutions for evil, once materialized. For a number of reasons, that is the wrong track in relation to climate change. We cannot wait until the planet has become unlivable. We must avoid the looming catastrophes to the extent possible. Stias, an independent research institute attached to the University of Stellenbosch (South Africa), has launched an international project to deal with a kaleidoscope of legal issues to the effect mentioned above.

Insofar as civil law can contribute to the avoidance of climate change, the means to arrive there need attention. Prof. Ulrich Magnus (before his retirement from University of Hamburg), one of the éminences grises of private and private international law in the German-speaking world, addresses a series of important procedural aspects of injunctive relief. Besides, he proposes and discusses the ins and outs of an International Environmental Court.

Criminal law could play an important role in arriving at a change of mindset. If irresponsible environmental behaviour would fall under the umbrella of criminal law, the deterrent impact would probably be significant. Prof. Gerhard Kemp, an eminent South African expert of criminal law, provides an overview of and develops useful thoughts about the potential of international criminal law as a key piece of global governance.

Africa is one of the continents that will suffer most from climate change. The legal scene of this ever more important continent is still largely unexplored. Prof. Oliver Ruppel, one of the key players in the arena of climate change, paints a vivid picture of major legal developments in Africa.

At the end of the day, legal action requires a better understanding of the do's and don'ts. To the extent possible, we must figure out what national states and enterprises have to do to meet their legal obligations. That issue, and substantive key topics in relation to injunctive relief, are discussed in my contribution.
A major part of this research project has been executed at the premises of Stias in Stellenbosch, a paradise for research. We are most thankful to Stias' director Hendrik Geyer and the extremely helpful Stias staff for their support and the welcoming and stimulating environment for groundbreaking research.

Jaap Spier

1 Injunctive Relief: Opportunities and Challenges: Thoughts About a Potentially Promising Legal Vehicle to Stem the Tide

Jaap Spier

> *Without addressing the interplay between climate change and justice, no response to climate change will succeed.*[1]

1.1 Introduction

1.1.1 High Noon

This book is the second fruit of a project conducted at Stias about the role the law could play to come to grips with the major challenges of our time, with emphasis on climate change.[2] If society at large does not change course drastically, we are heading towards global devastation. So far, there is very little reason to believe that politicians, business leaders and the public at large are going to embark on the bitterly needed reduction of Greenhouse Gas (GHG)[3] emissions.[4] On the contrary, two decades after 'Kyoto', global emissions are still increasing; see paragraph 1.2.

---

1 S. Johnson, 'Climate Change and Global Justice: Crafting Fair Solutions for Nations and Peoples', *Harvard Environmental Law Review*, Vol. 33, 2009, p. 301.

2 The first book, *Shaping the Law for Global Crises,* was written by Jaap Spier (hereinafter *Shaping the Law for Global Crises or Spier 2012*). It dealt with the major challenges of our time with a common denominator (climate change, unsustainable development, financial crises): irresponsible human behaviour with a huge adverse global impact; the book put emphasis on climate change.

3 *I.e.* carbon dioxide, nitrous oxide, methane, perfluorocarbons, hydrofluorocarbons and sulphur hexafluoride; *see* Parker & Blodgett, Greenhouse Gas Emissions: Perspectives on the Top 20 Emitters and Developed Versus Developing Nations, CRS Report for Congress, RL32721, 7 January, 2010; drafted by Larry Parker and John Blodgett footnote 7.

4 Others are a bit more optimistic. They believe that litigation may bring politicians to their senses. *E.g.* R.E. Blomquist, 'Comparative Climate Change Torts', *Val. U.L.Rev.* Vol. 46, 2012, pp. 1053-1055, with further references. But they acknowledge that it will take time; time we do no longer have if we want to avoid catastrophe. This is not to say that there are no promising developments. By way of example, two US Senators (Sanders and Boxer) have introduced a bill to reduce US greenhouse gas emissions. The bill aims at quite significant reductions by various means; *The Guardian*, 16 February 2013, available at <www.guardian.co.uk/commentisfree/2013/feb/16/bernie-sanders-climate-change-legislation>. It can only be hoped that their

True, there is an emerging sense of urgency. But the willingness to translate this into action is remote for various reasons.[5] To mention a few, focus on the short term (the *après moi le deluge* paradigm), the level playing field (quite a few important actors may be willing to change course, but not if they put themselves at a competitive disadvantage), pressure by interest groups, the idea that the cost of taking preventive measures outweighs the future benefits, wooden-headedness and arguably a kind of acceptance (a feeling that overcoming the deadlocked situation is a Herculean task, so 'we' should accept our fate).[6] And, last but not least, most of us do not feel personally affected.

If we were only frittering away our own future and would not do a lot of serious harm to others (many of them innocent victims), this stance would be difficult to understand, but it would arguably be acceptable. One is free to harm oneself. But we should and actually do know better. We are in the process of ruining our planet to the detriment of future generations and almost certainly also of the younger generation. We do cause a great deal of damage to citizens of very poor countries; countries that did not and still do not, or only marginally, contribute to the mess. Under these circumstances, a laissez-faire attitude is morally reprehensible. It is also wrong from a legal angle.

1.1.2 The Prospects of Legal Strategies

I realize that one should not overstretch the arm of the law. The challenges ahead of us are so colossal that there is no single solution (except wise and courageous politicians around the globe, but that is a mirage). Legal solutions will not rescue our planet, but they could

attempt will bear fruit. On 11 December 2012, the Attorneys General of seven US States have given notice to EPA of their intent to sue EPA for its failure to review the standards of performance for oil and gas operations and the failure to complete the emission guidelines for existing sources. According to Michael Gerrard, the US 'administration' could do quite a lot to stem the tide, even without Congress: M.B. Gerrard, 'Federal Executive Actions to Combat Climate Change', *The New York Law Journal*, Vol. 249, No. 49, 2013.

5 This is not to say that nothing happens. On 25 April 2012 Ceres (a group of major investors) has issued a press release about a "New Ceres/Sustainalytics Report". It observes "individual examples of leadership" in – *inter alia* – the field of climate change, but it also notes that "far too many companies are only taking small, incremental steps." There are some hopeful developments in the US too, and more could be achieved quite easily; *see* Gerrard, 2013. More generally, a series of reporting and responsible investment activities have emerged; *see*, by way of example, 'Principles of Responsible Investment, Integrated Analysis, How Investors Are Addressing Environmental, Social and Governance Factors in Fundamental Equity Valuation', available at <www.unpri.org/publications>; Carbon Tracker-Initiative, available at <www.carbontracker.org/wastecapital> and 2o Investing Initiative; *see also* the 'Six Prinicples for Responsible Investment', available at <www.unpri.org>. Ever more countries encourage the use or production of bio-fuels. Despite the laudable intentions, it is very much open to debate whether this strategy is a blessing for society. The impact on climate change is limited, whereas it jeopardizes poor people as basic grains have become more expensive: D. Capice, 'The Looming Food Crisis', 2011, *YaleGlobal online*, available at <http://yaleglobal.yale.edu/print/6855>.

6 For practical purposes, it means that we leave the mess to others who get the fate passed on.

play a not unimportant role if they would be sufficiently credible. That is what we aim to provide: legal solutions based on an underpinning as solid as possible.[7]
Why, then, could legal strategies potentially serve as an instrument to stem the tide? The answer is that in major parts of the world – particularly in the 'developed' world – enterprises and even states will comply with legal obligations. They will – perhaps grumbling, but still – comply with judgments. Quite a few key players (politicians and business leaders) will even cheer legal solutions as they would serve as convincing arguments to explain to voters and shareholders that they do not have a choice but to comply. Others may not be overjoyed that they are under a legal obligation to curb GHG emissions quite drastically, but they will at least appreciate that they are taught what they have to do.[8] The latter is also important for, *e.g.* auditors; without specific knowledge on this point, they are in the dark when drafting their opinions about annual reports by enterprises.[9]
Those who believe that 'their' courts will resist the temptation to come to the aid of victims of climate change, *i.e.* that they don't have to worry, may be mistaken. They overlook that judgments of groundbreaking courts can often be enforced in other countries where enterprises do have assets.[10] It may be an incremental process and it may take too long.[11] But as things stand, 'courting' climate change is, next to technical solutions and pressure from major stakeholders and the press, one of the more promising avenues to get at concrete results.

7 I will confine myself to referring to a few leading sources. A lot has been written about climate change; a great many publications are immensely valuable and often groundbreaking. So far, quite some emphasis is put on international law, human rights law, environmental law, constitutional law and the public trust doctrine. Relatively little is written about the potentially more promising angle of tort law; *see* about the latter topic D. Hunter & J. Salzman, 'Negligence in the Air: The Duty of Care in Climate Change Litigation', *University of Pennsylvania Law Review,* Vol. 155, 2007, p. 1741. To avoid misunderstanding, the other sources of the law are and remain important, if not for other reasons because they make it virtually impossible for national legislators to 'overrule' case law based on (as the case may be: bold) interpretation of well-established concepts.

8 A similar point is made by J. Peel, L. Godden & R. J. Keenan, *Transnational Environmental Law*, Vol. 1, 2012, p. 270.

9 *See* in more detail Spier, 2012, pp. 217 *et seq.*

10 As explained in Chapter 2, judgments of foreign courts against other States cannot be enforced in the defendant State.

11 Justice Brian J. Preston is probably right that "the judiciary, particularly at national levels, is [...] faced with the task of explicating the law of sustainable development, case by case. Incrementally, a body of environmental jurisprudence will emerge. In performing that task, national judiciaries will be assisted by the exchange of judicial decisions, information and experience between jurisdictions. In this way, national judiciaries may benefit from each other's knowledge, experience and expertise." B.J. Preston, *The Role of the Judiciary in Promoting Sustainable Development: The Experience of Asia and the Pacific*, p. 1, available at <www.lawlink.nsw.gov.au/agdbase7wr/_assets/lec/m4203011721754/speech_10jan2006_preston.pdf>. Not all courts are independent; other courts face difficulties of a different nature. But it may be expected that even courts reluctant to show willingness to deal with major challenges will, at some stage, "slowly wake up to the reality that they have this critical role," as Patricia Kameri-Mbote and Collins Odote put it ('Courts as Champions of Sustainable Development: Lessons from East Africa', *Sustainable Development and Law*, Fall 2009, p. 37).

1.1.3 *Prevention Is the Keyword*

The authors of this book believe that prevention is the keyword. We simply cannot accept that catastrophe has to set in before the law can come into play.[12] In a UNEP Training Manual on International Environmental Law, prevention is labelled as one of the most commonly encountered concepts of international environmental law,[13] while it is advocated as 'a golden rule' both for the environment and for economic reasons.[14] Shelton and Kiss are entirely right that the principle of prevention will most likely necessitate injunctive relief where the threat of harm is imminent or harmful activity is going on.[15] So far this has mostly happened in relation to – comparatively speaking – nitty-gritty cases. For traditional lawyers, it may seem quite a challenge to use the same legal weapons in the fight against climate change. Thus far, not every lawyer is prepared to believe that the law encompasses large-scale misery. It may take some time – we cannot afford to lose, but still – but I am confident that proactive remedies will become the most effective way to stem the tide. For that reason this book puts quite some emphasis on injunctive relief. Criminal responses may also have quite some impact. Particularly the belief that one could be exposed to long time sentences may bring people to their senses.

Seen from a moral and legal angle, it would be weird and wrong to close our eyes to the immense and still largely avoidable human tragedies and to the damage to the economy with a foreseeable adverse impact, particularly on the already most vulnerable people. The gist of this book is that the law can come into play to avoid looming losses, if caused by wrongful human activities. The 'better safe than sorry' paradigm is swiftly

12 Prevention is far from a novelty in the legal arena. It was already acknowledged by International tribunals before World War II; *see* E. Rieter, *Preventing Irreparable Harm, Provisional Measures in International Human Rights Adjudication*, Intersentia, 2010, p. 14 and pp. 21-22 and pp. 38-39; *see* in the context of environmental damage pp. 20 *et seq.*; *see also* pp. 1088 *et seq.*; comparatively speaking – seen from a global angle – provisional measures are often issued in minor cases. Rowena Maguire and Bridget Lewis also point at the importance of prevention: R. Maguire & B. Lewis, 'The Influence of Justice Theories and International Climate Politics and Measures', *MqJICEL*, Vol. 8, No. 1, 2012, p. 25. Although a different issue, seen from a doctrinal angle, even a creditor is obliged to take reasonable steps to reduce or avoid looming losses, whereas the debtor is under an obligation to compensate the reasonable expenses incurred by the creditor. That goes for the law of torts, but equally for contract and sales law. *See*, by way of example, respectively Art. 2:104 Principles of European Tort Law, and Art. 163 Common European Sales Law. More generally speaking, prevention does not get enough attention. *See e.g. Johnston et al. v. NEI International Combustion Ltd* [2007] UKHL 39; in those instances, injunctive relief would have been a much better option.

13 UNEP Training Manual on International Environmental Law, p. 30.

14 UNEP Training Manual on International Environmental Law, pp. 32, 33 and 54. *See* for similar views D. Shelton & A. Kiss, *Judicial Handbook on Environmental Law*, UNEP, 2005, pp. 22 and 59, available at <www.unpei.org?PDF/.../Judial-Handbook-Environment-Law.pdf>.

15 Shelton & Kiss, 2005, p. 59.

gaining ground.[16] There is an emerging belief and understanding that otherwise we will be confronted with irreparable losses and prohibitive costs.[17]

1.1.4 Damages: The Wrong Way Forward

Unlike a growing number of academics and practitioners (who are seemingly attracted by the – in their view – low hanging fruit of claims for damages), I don't think that compensation of the colossal losses is the right answer. Strikingly, (some) insurers seem to believe that they can manage climate change-related claims.[18] I do not believe that damages are the right way ahead.[19] First and foremost, it will create even more problems than it solves, given the magnitude of the harm that will occur if we cannot stem the tide. After all, if enterprises and national States will have to compensate the losses, they will often be ruined by the heavy burden of damages. That, in turn, will affect a great many people: employees who will lose their jobs, the unemployed, people who are unable to work due to illness or handicaps, given that the social security benefits will get under severe strain, those dependent on pension funds whose assets will be greatly affected, but not in the least people in the 'developing' countries; the latter will be confronted with a collapse of the global economy, which will unavoidably have a major adverse impact on them too. Secondly, the first will get all. That,

16 *See*, by way of example, C. von Bar 'Art. 1:102', *Principles of European Law, Non-Contractual Liability Arising out of Damage Caused to Another (PEL Liab. Dam.)*, Oxford University Press 2009. (Where a legally relevant damage is impending, this Book [sic] confers on a person who would suffer the damage a right to prevent it. The right is against the person who would be accountable for the causation of the damage if occurred.) *See also* Art. 6:301. According to the Commentary (Von Bar, 2009), prevention is "regarded as better than atonement for damage" (pp. 265 and 991). Von Bar adds that all legal systems of the EU recognize the right to demand of a person who is threatening to cause legally relevant damage to desist from his dangerous conduct (p. 269).

17 UNEP Training Manual on International Environmental Law, p. 33.

18 *See* the book review of C.M. Carroll *et al.*, *Climate Change and Insurance*, by M. Orie and W.R. Stahel, The Geneva Convention, Risk Management Newsletter No. 52, The Geneva Association, November 2012, pp. 21 and 22; *see* in more detail Spier, 2012, pp. 153 *et seq.* Carroll's book – and quite a few other insurance studies – tend to focus on the US perspective and litigation in the US, litigation that has not been extremely helpful to come to terms with the threats of climate change. True, there have been a few groundbreaking judgments, but most have been reversed in appeal. If not for other reasons, I wonder whether the focus on US litigation provides us sufficient basis for predictions about litigation in other parts of the globe. More importantly, insurers tend to – and almost necessarily have to – rely on historic data. In the context of climate change, that seems a very serious mistake. After all, leading experts are ever more pessimistic about the magnitude of the harm we are going to face if we stick to business as usual; despite some progress in the international negotiations and the willingness of a growing number of enterprises to curb their GHG-emissions (though rarely to the extent needed), global GHG emissions are still rapidly increasing. There is very little basis for hope that this is going to change in the near future; in turn, we notice a sharp increase of unusual natural events (catastrophes). This is only the beginning. Besides, ever more 'developing' countries are considering starting litigation about the cost of mitigation and adaptation; activist lawyers enter the scene. It follows that the odds are very much against the non-believers or those who believe that we can keep things under control.

19 *See* for legal instruments to avoid crushing liability (under English law) G. Kaminskaite-Salters, *Climate Change Liability under English Law*, Universitaire Pers Maastricht, 2010, pp. 95 *et seq.* According to Kosolapova damages is a non-issue for several reasons (Kosolapova, 2013, pp. 43).

however, would be very unfair to future victims whose losses will greatly exceed those of the victims in the near future. After all, according to the prevailing view, the situation will get ever worse, and by the same token the losses will get ever more colossal. Last but not least, those who will have to settle the bill will often not be responsible for the mess; they may have harvested or have inherited some fruits of our unsustainable lifestyle, but they will equally be confronted with the nasty consequences of climate change. I realize, of course, that this is harsh for victims. But it would be a serious mistake to focus only on 'direct' victims.

More importantly, our goal should be to avoid unnecessary human suffering, particularly so if it would affect millions of people around the globe, and our focus should be prevention.

The above leaves unanswered the question of how to cope with the major adaptation cost[20] and the cost of mitigation to be incurred by developing countries unable to bear these costs themselves. At least in relation to the latter, political compromises may well be within reach, as most politicians do understand that the very adverse consequences of climate change can be avoided only if almost all nations cooperate in reducing GHG emissions. All the more so as the global mitigation costs are manageable and reasonably affordable.

This leaves untouched, of course, that the threat of legal action and particularly liability for damages may have a preventive effect. Koivurova, quoting Brunnée, may well be right that

> Publishing their legal claims allows small states and indigenous people to reinforce their activities in the climate regime and obtain a louder win in the global regime.[21]

I have great sympathy for the argument, but for the reasons just mentioned I am afraid that it would be fraught with colossal risk if courts would indeed grant damage awards. That is by no means the predominant view. Damages belong to one of the foremost paradigms of lawyers. It is emphasized in the widely acclaimed Ruggie Principles[22] and also advocated by the World Bank.[23] In the overwhelming majority of cases, damages are only just and fair if a wrongful act or omissions could not be avoided. But in a few instances, particularly in the realm of climate change, it does more evil than good.

1.1.5 Injunctive Relief Just a Mirage?

The proof of the cake is in the eating. I am well aware of the fact that many of my submissions are novelties, largely borrowed from or based on well-established legal concepts and

20 *See* the impressive studies in M.B. Gerrard & K. Fischer Kuh (Eds.), *The Law of Adaptation to Climate Change, U.S. and International Aspects*, American Bar Association, 2012.

21 T. Koivurova, 'International Legal Avenues to Address the Plight of Victims of Climate Change: Problems and Prospects', *Journal of Environmental Law and Litigation*, Vol. 22, 2007, p. 298.

22 *See* in more detail Spier, 2012, pp. 37 *et seq*.

23 *See* World Bank Group, International Finance Corporation, *IFC Performance Standards on Environmental and Social Sustainability*. Effective 1 January 2012, p. 6. The – no doubt not anticipated – unsatisfactory consequence would be that 'developing' countries (by and large the 'clients' of the Bank) would have to pay compensation to the rich countries.

case law, be it in a rather different setting. So, in a sense, I build on unsafe ground. There certainly are quite a few counterarguments for almost all of my submissions.[24] Traditional – or should we say legalist? – courts will be keen to rely on these counterarguments as an 'excuse' to refrain from groundbreaking judgments.

As a matter of fact, the future of mankind lies to some extent in the hands of judges, willing to deliver bold judgments if and to the extent that States and enterprises are not prepared to meet their legal (and at the same time also moral) obligations.[25] We can only keep our fingers crossed that prospective victims will timely – *i.e.* in the very short term – get enough courts on board to effect far-reaching reductions of GHG emissions.

1.2 The 'Realities' of Climate Change

> Negotiations on climate change have been unsuccessful partly because they do not carry the sense of impending crisis that economic negotiations like the fiscal cliff talks in Washington do. Meanwhile the planet is inching close to a climate cliff.[26]

According to the predominant view among climate change experts, our globe is heading towards colossal harm due to global warming if we stick to business as usual and, more precisely, if a fatal tipping point of a global rise of temperature over 2°C, compared with pre-industrial times, is passed. According to recent research, the 2°C no longer constitutes the threshold of 'dangerous' climate change; instead, it is said to be extremely dangerous.[27] GHG emissions are at least a major factor in the rising of global temperatures.

Despite a great many pledges and lip service paid to the urgent need to achieve significant reductions of global GHG emissions, the aggregate global emissions are still increasing, partly due to the staggering economic progress in a few important and major emerging economies such as Brazil, India and Indonesia and the new economic giant China. The IEA's chief economist Fatih Birol warned that this trend is "perfectly in line with a

24 *See* in much detail D.A. Kysar, 'What Climate Change Can Do about Tort Law', *Environmental Law*, Vol. 41, No. 1, 2011, available at <http://ssrn.com/abstract=1645871>.

25 *See* for a similar view, among many others, M.C. Wood, 'Atmospheric Trust Litigation', chapter in *Climate Change Reader*, available at <www.law.oregon.edu/assets/.../mwood/atmo.pdf>, p. 37. The trust-idea undoubtedly is an interesting concept, but – as Wood readily admits – it is not apt to get rid of the "myriad of issues [that] may bar recovery" (p. 50). She rightly points at declaratory relief (pp. 51 *et seq.*), which – I think – largely faces the same obstacles as injunctive relief.

26 S. Barrett, 'All-Night Negotiations in Doha Produce No Cliffhanger', *Yale Global Online*, <http://yaleglobal.yale.edu/print/8277>.

27 K. Anderson & A. Bows, referred to in 'James Hansen and the Climate-Change Exit Strategy: Monthly Review', 2013, <http://monthlyreview.org/2013/02/01/james-hansen-and-the-climate-change-exit-strategy>.

temperature increase of 6°C (by 2050), which would have devastating consequences for the planet."[28]

On 19 November 2012, the World Bank issued an alarming report: Turn Down the Heat.[29] It paints a rather gloomy picture of the world that future generations will inherit if we stick to business as usual. The report serves

> as a rigorous attempt to outline a range of risks, focusing on developing countries and especially the poor. A 4°C world would be one of unprecedented heat waves, severe droughts, and major floods in many regions, with serious impacts on ecosystems and associated services. But with action, a 4°C world can be avoided and we can likely hold warming below 2°C.
>
> [...] Even with the current mitigation commitments and pledges fully implemented, there is a roughly 20 percent likelihood of exceeding 4°C by 2100. If they are not met, a warming of 4°C could occur as early as the 2060s.[30]

It can only be applauded that the World Bank sounds the alarm. But the message still seems overly optimistic, for understandable reasons. The World Bank is probably not in a position – or takes the view that it would be counterproductive – to depart from the most realistic scenario: global GHG emissions are still going to *increase* for quite some time. I am not an expert in the field of climate change science, but it seems quite likely that the latter – realistic – scenario implies that even more harm will materialize. It underscores the urgent need to change course and to be as creative as possible to find potentially promising avenues to get there.

Back to the impact of climate change. Once global temperature has increased with more than 2°C, a point of no return is reached.[31] That is far from an activist view. According to the German Advisory Council on Global Change

> [t]he vast majority of scientists now agree that if global warming exceeds a mean temperature of 2°C it will lead to dangerous, irreversible and practically uncontrollable consequences for both nature and mankind.

This view is endorsed by 133 countries.[32]

---

28 <www.reuters.com/print?aid=USBRE84NOMJ20120524>. It is all the more striking – and alarming – that *e.g.* the Netherlands Planning Office for the Living Environment (Leefomgeving) takes the view that the Netherlands is reasonably prepared for climate change; *see* its report issued 25 July 2012, *Effecten van klimaatverandering in Nederland*, 2012.

29 World Bank, *Turn Down the Heat, Why a 4°C Warmer World Must Be Avoided*, 2012, available at <www.worldbank.org>.

30 World Bank, 2012, Executive summary, p. 1.

31 At the Copenhagen Conference, the aim was still to keep an increase of GHG emissions below 2°C; <https://en.cop.15/dk/files/pd/copenhagen_accord.pdf>.

32 WBGU, *Solving the Climate Dilemma: The Budget Approach*, 2008, p. 1, available at <www.wbgu.de>. Further down on the same page the Council observes that "[n]ew research findings illustrate that the physical leeway for the protection of the Earth's climate has become very narrow. It is urgently necessary to take stock at both global and national levels."

If this 2-degree-threshold is passed, a series of adverse consequences will materialize. *Inter alia*, a significant sea level rise, in worst-case scenarios by several metres at the end of this century,[33] an increase of extreme weather events (including hurricanes, excessive rainfall and droughts), increased forest fires that burn hotter and are more destructive, shoreline erosion,[34] loss of genetic, species and ecosystem diversity, adverse impact on biodiversity, acidification of oceans, strain on water and food supply, health risks such as all kinds of diseases, a range of anxieties, depressive disorders and premature death,[35] environmental migration, not to speak of the more indirect impact on the economy and security.[36] The more temperatures will rise, the gloomier the picture will become. After all, damage will exponentially increase. New and even more fatal tipping points will be passed.[37] As so often, the already most vulnerable countries and people will be hit most.

It is ever more open to debate whether we are still able to avoid crossing of the 2°C threshold.[38] After all, the situation continues to deteriorate. The global emissions are still rising, in some countries at a staggering rate. More alarmingly, there is hardly any basis for hope that politicians will get the situation under control.

Even if we are able to stem the tide and to remain within the 2°C boundaries, the adverse impact of global warming will take a toll,[39] according to some experts even a dangerous one.[40] Most telling arguably is that a number of Small Island States recognize that it may be necessary to evacuate their people. The Maldives is already in the process of establishing a national trust fund for relocation to a new homeland.[41]

The prevailing view is not undisputed.[42] The critics adhere to diverging views. These views can be summarized as follows:

- Climate change is a phantom;
- It may be real, but belongs to the normal weather patterns;
- It *may* be real and even caused by GHG emissions, but we cannot (yet) draw firm conclusions;

33 J. Hansen *et al.*, *Scientific Case for Avoiding Dangerous Climate Change to Protect Young People and Nature*, 2013, p. 13, available at <http://pubs.giss.nasa.gov/authors/pkharecha.html>.

34 *Ibid.*

35 *Ibid.*, p. 15.

36 *See*, also for further references, WGBU-report, pp. 11-13.

37 *See* J. Hansen, 'A New Age of Risk, Mobilizing Science for Sustainable Development: The Sustainable Development Solutions Network', presentation at Columbia University on 22 September 2012.

38 The World Bank Report *Turn Down the Heat* also seems rather pessimistic, although its message is cast in diplomatic language.

39 *See* in more detail WBGU-report, p. 11 and for further references Spier, 2012, pp. 17-19; World Bank, 2012, p. 2.

40 Hansen *et al.*, 2013, p. 12.

41 J.H. Knox, 'Linking Human Rights and Climate Change at the United Nations', *Harvard Environmental Law Review*, Vol. 33, 2009, p. 498, also for further references.

42 *See*, by way of example, C.D. Idso, *The State of Earth's Terrestrial Biosphere*, Center for the Study of Carbon Dioxide and Global Change, available at <www.co2science.org> (5 December 2012); B. Wiskel, *The Emperor's New Climate*, 2006, Evergreen Environmental Company; V. Klaus, *Blue Planet in Green Shackles, What Is Endagered: Climate or Freedom?*, 2008, Competitive Enterprise Institute.

- We still have enough time to cope with the adverse consequences; or
- We can reasonably rely on scientific progress that will enable us (future generations) to come to grips with the threats.[43]

Criticism is often cast in rather strong language.[44] Some believe that the firm tone hides the weakness of the propositions. I do not think that we have to take a stance in this debate. Although I do not have any reason at all to believe that the majority view is incorrect, I reckon with the possibility that it is mistaken. We have learned from the past that majority views are not always right. Be that as it may, one can and should not lend one's ears to the sceptics. In that respect the precautionary principle paves the way. We simply cannot afford to take the risk that the sceptics are mistaken by refraining from taking bold steps to reduce GHG emissions. After all, if we would stick to business as usual and it would turn out that the sceptics were mistaken, unnecessary and irreparable colossal damage would be done to a great many people, future and present.

In his statement at the Doha conference in November 2012, the IPCC chairman Pachauri reminded his audience that he had asked the following rhetorical question at the acceptance of the Nobel Prize 2007: "Will those responsible for decisions in the field of climate change at the global level listen to the voice of science and knowledge, which is now loud and clear?"[45] The question was perhaps not so rhetorical as the answer was yes and no at the same time. Most of the people in responsible positions suggest that they do listen, thus paying lip service to the urgent need for global action. But more importantly, they do not translate their laudable words into (appropriate) action.[46] Hence the justification for a discussion of injunctive relief.

1.3 Causation

1.3.1 Introduction

Climate change is different compared with any other issue humanity has faced so far. According to the prevailing view, climate change is brought about by a series of factors.

43 That seems by and large the US position; *see* CSR-report for Congress, Greenhouse Gas Emission Drivers: Population, Economic Development and Growth, and Energy Use, 5 March 2010, No. RL33970, drafted by Blodgett & Parker, 2010, pp. 31 and 32, hereinafter referred to as CRS-report 2. The report rightly poses the question whether "incremental 'muddling through' " really is an option. It is equally right in pointing at the "unpredictable breakthroughs" (p. 32), but that said, the report falls silent (undoubtedly because the authors could not openly take a position).

44 *See* for examples Spier, 2012, pp. 14 *et seq.*

45 Statement of 28 November 2012; can be retrieved from the IPCC website. According to Lawrence M. Krauss, "ideological biases have become so ingrained in Washington that scientific realities are subordinated to political intransigence" (L.M. Krauss, 'Deafness at Doomsday', *The New York Times*, 15 January 2013, quoted from *Yale Global Online*). This regrettable state of affairs is probably not confined to the US, although rather extreme right-wing forces in that country are certainly unhelpful to get at meaningful action.

46 According to Scott Barrett, an important factor is that "the cliff is uncertain," or "that there is no reason to believe that 2°C really is the edge of the cliff *et seq.*" Under those circumstances, negotiations often prove ineffective.

Human activities have played and still play an important role; at least, that is the predominant view. I have little doubt that few courts will deny that, according to the state of the art (climate change science), there is a causal relationship between human activities (GHG emissions) and climate change. After all, leading experts – by no means only the authoritative IPCC – take it that this can hardly be disputed any longer. *Full certainty* is generally not required, as judgments in a series of cases – such as tobacco, asbestos and medical malpractice – have shown in many countries. This does not mean that judges discredit the non-believers; as mentioned in paragraph 1.2, they may be right. But it would be fundamentally unfair to victims – in our scenario to society at large – to wait until all experts have arrived at the same conclusion, if that is going to happen at all (climate change has become big business).[47] Thus we have only removed the easiest part of the causation problem, *i.e.* causation on a global level.

The difficulties lie in the complexity of the phenomenon. Up to a certain level of GHG emissions, the climate will not be greatly affected. As we have seen in paragraph 1.2, most scientists take the view that an increase in global temperature up to 2°C would give rise to difficulties on a global scale, but that the extent may be manageable. If that threshold is passed, a point of no return is reached and catastrophe will set in. The extent of very adverse consequences will, understandably, increase, the more GHG is emitted.

The change of the climate can largely be attributed to emissions in the past. Owing to the additional emissions in our era, the magnitude of the harm will increase substantially. The number of emitters is incalculable. A major part is brought about by a few countries and regions (particularly the so-called developed nations). But the global picture is changing quite rapidly. Cumulatively, from 1850 to 2006, Annex I countries to the Kyoto Protocol (largely the developed countries)[48] had emitted approximately 74% of energy-related CO_2. In 1990, Annex I countries accounted for 60% of emissions of GHG. By 2005 the figure has dropped to 47%.[49] By 2005, the top twenty emitters (no longer only the 'developed' nations) accounted for 75% of global emissions.[50]

GHG emissions stem from all kinds of activities: commercial and private (*i.e.* the 'ordinary' people). States, too, do emit GHG (governmental activities and enterprises owned by the government).

By and large, we do understand the contributions from the past and the present. We can calculate the emissions per enterprise and per national State fairly accurately. According to the President of the (US) National Academy of Sciences – in a testimony before the US House of Representatives

47 Kaminskaite-Salters, 2010, is more cautious (p. 169), but in my view overly cautious.

48 With the exception of Japan and Turkey (if counted as an Asian country), no Latin American, African or Asian country belongs to the Annex I countries.

49 CRS-report, p. 2. These recent figures are worrying as they show that the extent of reductions bitterly needed cannot be achieved by the Annex I countries only, leaving aside the moral and legal aspect. If land use would be included, the percentage would be 41.2 (p. 8).

50 CRS-report, p. 4. "Land use" and forestry are not taken into account in this figure.

> we understand the mechanisms of CO_2 and climate better than we do of what causes lung cancer. [...] In fact, it is fair to say that global warming may be the most carefully and fully studied scientific topic in human history.[51]

Will that do for injunctive relief? Many eminent lawyers answer the question in the negative, or perceive causation as one of the major stumbling blocks,[52] primarily because the GHG emissions of the respective countries and enterprises are small, with a few exceptions as we will see below.

Although those who believe that causation will turn out to be an insurmountable hurdle[53] unquestionably have a point, I am inclined to believe that they *may* be unnecessarily pessimistic. Under Japanese law, for example, the chances are said to be quite favourable if unlawfulness can be established.[54]

1.3.2 The Burden of Proof

In many legal systems, the plaintiff will generally have to prove causation; that also goes for international law.[55] Blomquist points at Chinese law – environmental torts – that would depart from "the principle of causation presumption," *i.e.* it would reverse the burden of proof ("the party causing the injury shall bear the burden of proving that there is no causality between the act causing the injury and the harmful result").[56] I wonder whether

51 Quoted by Kysar, 2011, pp. 29 and 30.

52 *See, inter alios,* most country reports *in* R. Lord *et al., Climate Change Liability, Transnational Law and Practice*, Cambridge University Press, 2012, particularly the Overview, p. 33; the Australian report by R. Abbs, P. Cashman & T. Stephens, p. 92; the Chinese report by Deng Haifeng, p. 137; the German report by H.-J. Koch, M. Lührs & R. Verheyen, p. 400; the Indonesian report by M.A. Santosa, J. Khatarina & R. Sjarief Assegaf, p. 197; the Israeli report by I. Rosen-Zvi, p. 291; the Japanese report by Y. Takamura, p. 229; the Kenyan report by P. Kameri-Mbote & C. Odote, p. 315; the Mexican report by J.J. González Marquez, p. 639; and the South African report by J. Glazewski & D. Collier, p. 340, be it that the observation is particularly made in relation to the link between a specific loss and a specific event. D.A. Grossman, 'Warming Up to a Not-So-Radical Idea: Tort-Based Climate Change Litigation', *Columbia Journal of Environmental Law*, Vol. 28, No. 1, 2003, pp. 58 and 59, is more optimistic, at least "during the continued progress of global warming."

53 *E.g.* R.F. Blomquist, 'Comparative Climate Change Torts', 46 *Val. U.L.Rev.* 2012, p. 1053, at p. 1057; Kysar, 2011, p. 54. Hunter and Salzman quote – *inter alia* – a judgement by the US District Court of Southern Mississippi pointing at "daunting evidentiary problems" in the arena of causation: Hunter & Salzman, 2007, p. 1744.

54 Blomquist, 2012, p. 1068, although this does not necessarily follow from the quotation. It is said to be an obstacle under South African law: idem, p. 1071.

55 D. Shelton, *Remedies in International Law* (2nd edn), Oxford University Press, 2005, p. 317.

56 Blomquist, 2012, pp. 1065 and 1066. This certainly is an interesting concept as it is very plaintiff friendly. In many legal systems and also in international law, the burden of proof lies upon the party who submits specific facts. As the ICJ put it: "the Court considers that, in accordance with the well-established principle of *onus probandi incumbit actori*, it is the duty of the party which asserts certain facts to establish the existence of such facts. This principle which has been consistently upheld by the Court applies to the assertions of facts by the Applicant and the Respondent. The Court observes that it is of course to be expected that the Applicant should, in the first instance, submit the relevant evidence to substantiate its claims. This does not,

this submission will overcome the causation hurdle. Even in this reasoning, the question remains whether a very, very small contribution can be labelled as a legal cause.

1.3.3 Acts or Omissions?

Injunctive relief presupposes a threat of an illegal act or omission.[57] GHG emissions that go beyond the legally acceptable limits can be cast in terms of both acts and omissions. The more attractive view probably is to perceive them as acts (the continuing excessive[58] emissions), but one could also argue that the 'wrong' lies in failing to reduce the emissions. For the purpose of this chapter this distinction is of little relevance.

1.3.4 Are Damages and Injunctive Relief Governed by the Same Causation Criteria?

Before discussing the taunting issue of causation in relation to injunctive relief, one should bear in mind that the causation doctrine has largely been developed for the purpose of claims for damages. Doctrine and case law are far less settled in relation to injunctive relief.[59] Causation also comes into play in relation to injunctive relief, I think. After all, injunctive relief requires a threat of damage. If a specific act or omission could not be considered as a cause of the loss, there is no (legally relevant) threat. Various courts and academics take the view that the requirements for injunctive relief are less strict than those for damages.[60] The US Supreme Court has taken a different stance.[61]

Be that as it may, we may fairly assume that the prevailing doctrine, developed in the field of damages, equally applies to injunctive relief if the GHG emissions of a specific defendant exceed the level of 50%. In many legal systems the causation hurdle is taken once that threshold is passed.

---

however, mean that the Respondent should not co-operate in the provision of such evidence as may be in its possession that could assist the Court in resolving the dispute submitted to it" (*Case concerning Pulp Mills on the River Uruguay* (*Argentina v. Uruguay*), Judgment of 20 April 2010, pp. 17 and 18.

57 That may be an obstacle in various jurisdictions; *see e.g.* Abbs *et al.*, 'Australian Report', *in* Lord et al., 2012, p. 94. The issue will be discussed in para. 1.4 below.

58 *I.e.* beyond the level legally acceptable.

59 It is hardly addressed in the context of injunctive relief in Lord *et al.*, 2012 and most other important studies about legal remedies in the realm of climate change.

60 *See e.g.* K.H. Engel, 'Damages, Injunctions, and Climate Justice: A Reply to Jonathan Zasloff', 58 *UCLA Law Review Discourse* 2011, p. 189, at pp. 202 *et seq.*, discussing a few nineteenth century cases in the realm of public nuisance and referring to other academic writings. He suggests the application of the same reasoning to tort law; if I understand correctly, this is also Roda Verheyen's view: R. Verheyen, *Climate Change Damage and International Law, Prevention Duties and State Responsibility,* Leiden, Boston, Martinus Nijhoff Publishers, 2005, p. 252. *See also NJDEP v. Dimant*, New Jersey Supreme Court, 26 September 2012; *Raymond E. Britt v. The Cyril Bath Company et al.*, 417 F.2d 433.

61 Los Angeles County, *California v. Humphries et al.*, 562 U.S., 2010, be it in the context of a specific legal provision. It seems to follow from the final paragraph that the court might have taken a different position if otherwise no redress would be possible.

Even in common law countries, a percentage of over 50% likelihood will suffice.[62] Prima facie that is not entirely true. After all, the more-likely-than-not doctrine is about something else: *i.e.* the likelihood of a causal link between a *specific* event and a *specific* damage. That, however, is not the issue dealt with in this chapter. In the cases discussed here, we do. If the likelihood that a specific defendant has caused the loss would exceed 50%, causation would not be an obstacle. Climate change is not a matter of likelihood; the contribution of a specific defendent is known. The difficulty is that the contribution of no single defendant exceeds 50%, neither in relation to the global losses nor in relation to any specific loss.
Claims for damages may stand better chances than injunctive relief if we ignore causation.[63] Particularly, the political argument, to be discussed below in paragraph 1.9, may be an insurmountable hurdle in relation to injunctive relief and could arguably be removed if damages are sought. It would be very unsatisfactory if the latter would hold true, as it would mean that catastrophe has to set in first before the law can come into play. That may, or may not, be inevitable and acceptable in the run-of-the-mill cases; it does not appeal if colossal and grievous harm to a great many people is at stake.

1.3.5 A Global Perspective Suffices?

It follows from paragraph 1.2 that climate change and, by the same token, GHG emissions pose a very significant threat to mankind. Does this finding suffice for injunctive relief against each and every State and enterprise whose emissions exceed the level legally allowed, even if the GHG emissions of a specific enterprise or State are minimal? An answer in the affirmative would make the rest of this chapter superfluous.
Seen from a global perspective, all GHG emissions *together* cause climate change. If we could perceive causation from such a global angle, the contributions of respective players would not matter, not even if they would only have a very marginal, or immeasurable impact on the upcoming or threatening losses. This global approach may not appeal to every court, but this concept should not be labelled as unrealistic from scratch,[64] as the following example may show. Certain pollutants into the air from large combustion plants (such as sulphur dioxide) should not exceed a certain level. Otherwise they adversely affect health.[65] Let us assume, for the sake of argument, that the negative impact on the

62 *See, e.g.* K. Oliphant, *in* B. Winiger *et al.* (Eds.), *Digest of European Tort Law*, Vol. 1: Essential Cases on Natural Causation, 2011, pp. 570 *et seq.*; *see* for a recent Canadian judgment: *Spieser v. Canada (A.G.)*, 2012 QCCS 2801. Kadner-Graziano observes that "the lowest percentage of probability required in European tort law systems is 51%" (T. Kadner-Graziano, *in* Winiger *et al.* (Eds.), 2011, p. 586 and European Group on Tort Law, *Principles of European Tort Law, Text and Commentary* (Spier), 2005, pp. 57 and 58.

63 *See*, by way of example, M.B. Gerrard, 'What Litigation of a Climate Nuisance Might Look Like', *Yale Law Journal*, YLJO Essay, 13 September 2011.

64 Thus, also, Brunnée *et al.*, 'Overview of Legal Issues Relevant to Climate Change', *in* Lord *et al.*, 2012, p. 33. Kosolapova discusses a wealth of particularly Australian case law (Kosolapova, 2013, pp. 97 *et seq.* and 126).

65 *See inter alia* Directive 2001/80/EC of the European Parliament and of the Council of 23 October 2001, L 309/1.

health is rather severe, that a great many people will be affected and that there is a fair chance that the critical level will be passed due to the emissions of, say, 1,000 enterprises of various sizes. Let us also assume that each of these enterprises quite easily and at relatively modest cost could reduce the level of their sulphur dioxide emissions. Would injunctive relief against each and every single defendant founder on the relatively small contribution of each of them? Or could one, *for the purpose of injunctive relief*,[66] depart from a more global perspective, *i.e.* the *enterprises together* will bring about considerable harm to many people and that mere fact suffices for injunctive relief to ward off the materialization of the loss? Legal doctrine on this point is seemingly still in its infancy.[67]

Yet potential victims could borrow from a recent and groundbreaking judgment of the US Court of Appeals.[68] In that case, several States, the city of New York and three private land trusts sought injunctive relief requiring a decree setting certain limits upon carbon-dioxide emitters (four power plant companies and a fossil fuel-fired power plant). The claim was based on the submissions that climate change poses a series of serious dangers, while the defendants are the five largest emitters in the US with collective annual emissions constituting 25% of the emissions from the domestic electric power sector, 10% of the emissions of all domestic human activities and 2.5% of the emissions worldwide. The Court goes at length to discuss a series of issues. After having found that plaintiffs have sufficiently alleged current and future harm, it turns to causation (in the context of public nuisance):[69]

> To satisfy the causation requirement, the alleged injury must be "fairly traceable to the actions of the defendant." This requirement "ensures that there is a genuine nexus between a plaintiff's injury and a defendant's alleged conduct," and "is in large part designed to ensure that the injury complained of" is "not the result of the independent action of some third party not before the court."

The court then turns to the submissions of the parties. Defendants argued that the plaintiffs could not isolate which alleged harm will be caused by the defendants' emissions, nor that such emissions would alone cause future harm. The Court finds these arguments unconvincing as the alleged injuries "are 'fairly traceable' to the actions of the Defendants." It is equally not responsive to the argument that the traceability must be evaluated in accordance with the standard by which the common law public action imposes liability on

66 This may well be very different for damages.

67 Under Australian law plaintiffs are ordinarily required to "provide an undertaking as to damages in connection with the effects of any prohibitory order sought": Abbs *et al.*, *in* Lord *et al.* 2012, p. 101; Y. Takamura (Japan), *in* Lord *et al.*, 2012 pp. 231-232, be it with some caution; H.J. Koch *et al.*, Germany, *in* Lord *et al.*, 2012 p. 411; J. Glazewski & D. Collier (South Africa) according to the "traditional formulation", p. 341. According to Indonesian law 'nature' (*i.e.* weather, climate etc.) will probably be considered as the cause of the loss (M-A. Santosa *et al.*, *in* Lord *et al.*, 2012, p. 197).

68 *Connecticut et al. v. American Electric Power Company Inc et al.*, 21 September 2009; reversed by the US Supreme Court in a judgment dated 20 June 2011; the Supreme Court explicitly left this point open.

69 References and footnotes omitted.

contributors of indivisible harm, referring *inter alia* to para. 840E and 875 of the Restatement (Second) of Torts. It also refers to a judgment by the Powell Dyffryn Court:[70]

> [T]his likelihood may be established by showing that the defendant has (1) discharged some pollutant in concentrations greater than allowed by its permits (2) into a waterway in which the plaintiffs have an interest that is or may be adversely affected by the pollutant and that (3) this pollutant causes or contributes to the kinds of injuries alleged by the plaintiffs [. . .]

The Powell Duffryn Court observed that

> [i]**n order to obtain standing, plaintiffs need not to sue every discharger in one** action, since the pollution of any one may be shown to cause some part of the injury suffered. The size of the injury is not germane to standing analysis.[71]

The Court concludes that

> [i]n view of this widely accepted case law, and the procedural posture of the case, Defendants' argument that many others contribute to global warming in a variety of ways, and that therefore Plaintiffs cannot allege traceability, does not defeat the causation requirement.

This judgment shows that causation does not *need* to be an obstacle. The Court's view is certainly appealing, but it may not convince every other court around the globe.[72] After all, it remains quite a step to bypass the minimal causation defence.
To illustrate this point, another judgment of the US Court of Appeals may serve.[73] The Court held that "the standing requirement prevents the virtually limitless spread of judicial authority." It notes that "'causation' in this context is something of a term of art, taking into account not merely an estimate of effects but also considerations related to the constitutional separation of powers." It subsequently delves into the factors to be proven by the plaintiff: a substantial likelihood that the defendant's conduct caused the plaintiff's injury, in fact, and that the asserted injury does not result from the independent source of some third party. The causal line should not be "too attenuated." The second requirement comes down to the following: absent the challenged conduct, there should be a substantial probability that the plaintiff would not have suffered the alleged injury. Unlike the Court of First

70 *I.e.: Public Interest Research Group of N.J. v. Powell Duffryn Terminals Inc.*, 913F.2d at 64.
71 *Public Interest Research Group of N.J. v. Powell Duffryn Terminals Inc*, 913 F.2d at 72.
72 Nor would it be irrelevant (even) if it would have been reversed. After all, non-US courts are not bound by US case law. The gist of discussing this case lies in the merits of the arguments put forward by the Court of Appeals.
73 *Ned Comer et al. v. Murphy Oil et al.*, US Court of Appeals for the Fifth Circuit of May 7, 2010, 2009 U.S. App. LEXIS 22774. The Court refers to a wealth of other cases.

Instance, the Court of Appeals rejects a standard under which the traceability requirement is satisfied if the complaint alleges that the defendant's conduct merely 'contributes' to the types of injuries alleged, regardless of the role third parties may have played.
So we have to address 'minimal' causation in more detail. Before doing so, we have to look into a few preliminary issues.

1.3.6 *Is There a Sufficient Threat?*

A further preliminary question is about the threat. As discussed above, it is beyond cavil that humanity faces very serious threats. But are these threats sufficiently precise and well understood to justify injunctive relief? Again, an example may illustrate this point. Let us assume that the people of, say, Florida or Louisiana come to realize that their region has become more hurricane-prone than in the past.[74] Could the – probably justified[75] – fear of ever more serious hurricanes in, say, the next five years in these States serve as a sufficient *factual* basis for injunctive relief, leaving aside against whom? An answer in the affirmative presupposes that it is sufficiently clear that there is or may well be a causal link between the increasing number of severe hurricanes in the next five years in this specific region and climate change. Defendants will certainly deny such a link and it may be difficult to prove,[76] unless plaintiffs would base their request for injunctive relief on the *longer-term* impacts of climate change (say the threat of ever more climate change-related hurricanes in the next thirty years).[77]
This point by no means suggests that injunctive relief is doomed to fail for lack of a sufficient factual basis. But prospective plaintiffs would be well advised to consider very carefully whether the factual foundation for their case is strong enough. Unlike, arguably, the just mentioned hypothetical case of the people of Florida and Louisiana, a case submitted by the people of the Small Island States or Bangladesh may stand a much better chance, at least as regards the factual basis of a serious threat in the somewhat longer term.

1.3.7 *Historical Contributions*

Back to the *legal* aspects of causation. First, the historical contributions.[78] They probably are a legal nightmare for claims for damages. In my view, emissions in the past can

74 According to the *Huffington Post* of November 9, recent polls show that global warming is a serious problem to 65% of Americans. *See* for earlier data: A. Leiserowitz, N. Smith & J.R. Marlon, *American's Knowledge of Climate Change*, Yale University, Yale Project on Climate Change Communication, 2010.

75 World Bank Report, *Convenient Solutions to an Inconvenient Truth* (report 51838), 2009, p. 1.

76 *See* for a similar view, *inter alios*, Koch, Luers & Verheyen, *in* Lord *et al.*, 2012, p. 403.

77 The IPCC findings may well suffice as a factual basis for such a submission. In case of class actions, the plaintiff(s) should not be required to show threat of damage in relation to each and every single member of the class. A judgment of the Supreme Court of California in *Re Tobacco II cases v. San Diego County* supports the latter submission, also discussing other cases.

78 In this context, data about land use and forestry are not available; CRS report, p. 6.

largely[79] be ignored in relation to injunctive relief. After all, that relief must be based on the submission that the *present* level of emissions is going to contribute to grievous damage to many people.

The following example may illustrate this point. Let us assume that the water of a river is safe for drinking purposes if the level of substance X remains below 100. In the past a series of factories (1-100) have discharged X in the river. At some point in time, the aggregate level of X in the river amounts to 90. At that stage, A gets the brilliant idea to build a new factory along the river's shore that will also emit X in the river. Together with the emissions of the other factories, the fatal threshold of 100 is nearing. Enterprise (B), which needs drinkable water for its production, and a few owners of houses built along the river, seek injunctive relief, *inter alia* from A. It is very unlikely that A can invoke the defence that he did not emit X in the past.[80]

1.3.8 The Condicio Sine Qua Non *Requirement*

Richard Wright wrote in a highly interesting article about causation:

> [. . .] I will only state my disagreement with the general view [. . .] that it is hopeless to try to articulate any definite, workable rules that limit the extent of legal responsibility for tortiously caused consequences, . . .[81]

Below, I am trying to find a legal basis for workable, fair and enforceable legal obligations on States and enterprises whose contributions to the global threat are marginal or at best small. That is a wrongfulness issue. It comes only into play if causation is not an obstacle.

Thus, we enter the arena of the condicio sine qua non requirement.[82] Each defendant could argue that its emissions do not contribute to, let alone will bring about, the changing

79 One could possibly argue that defendants that, compared with States or enterprises in similar circumstances, have emitted excessive amounts of GHG in the past would be a better target for injunctive relief than other prospective defendants. This submission presupposes that the 'excessive' emissions were unlawful on the basis of the law as it stood in the past. If that would be the case, the 'old emissions' could perhaps be added to the present ones for the purpose of determining the defendant's contribution in reducing the global emissions. Even in this reasoning, the contributions of most States and enterprises to the threat of global devastation remain marginal.

80 This leaves untouched that the enterprises that have brought about the historical contribution may be held liable for the losses caused by the discharges in the past and/or that their reductions should exceed those of A to avoid the passing of the threshold of 100.

81 R.W. Wright, 'Once More into the Bramble Bush: Duty, Causal Contribution, and the Extent of Legal Responsibility', *Vanderbilt Law Review*, Vol. 54, No. 3, 2001, pp. 1131-1132.

82 Given that this part of this book is about injunctive relief and not about damages, I do not have to address legal attribution (the question whether a specific defendant had to compensate a specific loss, once the condicio sine qua non-obstacle has been removed). Legal 'attribution' is a huge topic in its own right.

climate. After all, if its emissions were ignored, the same damage would occur or would have occurred. Seen from a *factual* angle and ignoring the top five emitters of GH gases, the defence is and is not entirely correct. As a matter of fact, the same would have happened, be it a bit later on in the time line (in relation to most defendants, the additional time required to get at the same global level of GHG emissions is hardly measurable). Some courts may accept this reasoning.[83]

Nonetheless, the defence seems invalid.[84] Once again the example mentioned in paragraph 1.3.7 may illustrate why. Each single factory could argue that the damage would materialize anyway, even if it had refrained from emitting substance X into the river. As a matter of *fact*, there would be some truth in that argument. But courts are unlikely to go along with the defence, if not for other reasons because it would mean that actual harm would have been caused by nobody. After all, if the defence were valid, it could be invoked by each of the factories.

1.3.9 Damage in a Specific Case

Unlike what applies to claims for damages, we can ignore the question whether a *specific* damaging event (say a hurricane) can, or cannot, be attributed to climate change.[85] After all, those kinds of questions arise after the event once a loss has materialized. Injunctive relief aims to prevent losses. Irrespective of the difficulty of demonstrating whether a causal link can be proven after the event, it is a shot in the dark before the event. It is unlikely that potential victims will base their submissions for injunctive relief on the threat of a very *specific* event, say a specific hurricane.

With the caveats discussed below, potential victims can base submissions for injunctive relief on the threat of *hurricanes in general, e.g.* the fear that their region has become more hurricane-prone, if there is sufficient scientific evidence for such a threat. The same goes, *a fortiori,* for countries such as the Maldives, the Small Island States and Bangladesh, to name just a few, where the threat of rising water levels is concerned.

83 *See, e.g.* D. Farouk & L. Youssef, *in* Lord *et al.*, 2012, p. 265; Shelton, 2005, p. 317.

84 *See*, also for further references, Spier, 2012, pp. 175-176.

85 *See* in more detail Kysar, 2011, pp. 30 *et seq*. According to Myles Allen, there is "in fact [...] no difference in principle between attributing causes for a single weather event and attributing causes for the observed trend in global mean temperature. In both cases, an event is observed [...] that could have occurred, at some level of probability, in a pristine climate," *in* Lord *et al.*, 2012, p. 12. This leaves untouched, I think, that it may be difficult to prove with a sufficient level of certainty, a causal relationship between a specific event (such as a flood or a hurricane) and climate change, particularly so for claims for damages. More likely than not, science will progress and the link may be proven at some stage. *See* about this issue also Grossman, 2003, pp. 22 *et seq.*; B.C. Mank, 'Civil Remedies', *in* M.B. Gerrard (Ed.), *Global Climate Change and U.S. Law*, 2007, pp. 200 *et seq.* and the in-depth analysis of Verheyen, 2005, pp. 248 *et seq.*

1.3.10 The Respective Players

The real obstacle for injunctive relief lies in the usually very, very small contribution of a specific enterprise or national State. We may have to distinguish between various scenarios
(a) national states;
(b) multinational enterprises;
(c) "local" enterprises;
(d) people generally.

As to national states, the level of GHG emissions diverges considerably. That having been said, it makes quite a difference whether one counts per capita or per country. If **counted per capita, a series of countries hardly emit any GHG: Afghanistan (0), Guinea,** Mozambique and Burkino Fasso (0.1), Bangladesh and Laos (0.3) belong to the low end of the spectrum. A few OPEC countries are at the other end: the United Arab Emirates (25), Kuwait (30.1) and Qatar (49.1). To the latter group also belongs Trinidad and Tobago (37.4). Most 'developed' countries belong to the wide range somewhere in the middle: Hong Kong[86] (5.5), France (5.9),[87] South Africa (8.9), Germany (9.6), the Netherlands (10.6), the Russian Federation (12.0), Canada (16.3), the United States (18) and Luxembourg (21.3). Palau (10.5) needs to be mentioned specifically as it tries to get green light for asking for an Advisory opinion of the ICJ. Counted per capita, the contributions by Indonesia (1.7) and Brazil (2.1) are modest.[88] If counted per country, by 2005, China, the United States, the Russian Federation, India, Japan, Brazil, Germany, Canada, the United Kingdom, Mexico, Indonesia, South Korea, Italy, Australia, Iran, France, Ukraine, Spain, South Africa and Turkey are the top twenty emitting countries (land use and forestry not included). Only the first eight of them are each responsible for 2% or more of the global GHG emissions.[89] If land use and forestry were included, Brazil and Indonesia would rank respectively third and fourth by 2005.[90]
For now, I resist the temptation to discuss whether or not one should count per capita. After all, that is not a causation – but instead a wrongfulness – issue.[91] It may suffice to mention that relatively modest emissions per capita should not necessarily mean that the country in question is not under a legal obligation to reduce its emissions, be it to a lesser extent than countries with a higher level of emissions per capita. There are mainly two reasons for this submission: otherwise, it will be close to impossible to arrive at the level of reductions needed

86 Part of the PR of China.
87 Probably due to major dependence on nuclear energy.
88 *See* <http://data.worldbank.org/indicator/EN.ATM.CO2E.PC>.
89 CRS report, Appendix B.
90 CSR report, p. 6.
91 *See* for an elaboration Spier, 2012, pp. 127 *et seq*. Interestingly, a report for the US Congress suggests "a flexible strategy that permits each country to play its strengths" (CRS report, p. 12), probably a nice way of putting that each country can opt for a 'strategy' that allows little advance.

to avoid devastation,[92] and even these countries usually have a group of well-to-do citizens; the latter could, at least, contribute to the good cause.[93] Below, I assume that the emissions of a specific country exceed the limits legally allowed.[94] If they do not, injunctive relief is doomed to founder from the beginning and the causation issue becomes irrelevant.

1.3.11 Injunctive Relief against Natural Persons and Small Enterprises?

For practical purposes one could prima facie ignore the class of natural persons. After all, no single person contributes to a measurable extent. More importantly, litigation conducted against natural persons would often be a waste of time, money and energy. Last but not least, it will not be easy to collect the relevant information about the emissions of specific private persons, unless they would be required to disclose relevant and often rather personal information.[95]

In an ideal world, *States* should assume their responsibility to curb GHG emissions within their territory to the extent needed.[96] If they do not, courts should be willing to step in and to urge them to comply with their obligations. Particularly in the short term, that is a rather optimistic scenario. Quite a few courts will shy away from bold judgments against their national States. Other courts are insufficiently independent; they have to sing for their supper. In those instances, litigation against enterprises could be a solution. But it will not bring enough solace. After all, the aggregate GHG emissions of enterprises are only part of the national emissions. So at some stage, it may turn out to be inevitable to consider litigation against natural persons if a State is unwilling to reduce the GHG emissions at a national level and injunctive relief against such a State is impossible.[97] This kind of litigation would primarily serve as a warning. It might have some impact if the 'right' targets are chosen.

Admittedly, injunctive relief against natural persons may be an unrealistic scenario too for at least three reasons. First, it is (very much) open to debate whether courts, unwilling to issue injunctive relief against their national states, would be willing to consider such relief against private persons. Secondly, it remains a time consuming and costly exercise. Thirdly, and most importantly, the contributions of natural persons are very, very small. It is only worth the effort if victory for the plaintiffs would bring other natural persons to their senses without further litigation.

The point made in relation to natural persons largely goes for small (national) enterprises such as grocery shops as well. It may serve a useful purpose if a specific small enterprise

92 Bearing in mind that modest per capita contributions of, say, India, Brazil and Indonesia end up in not insignificant aggregate emissions of the entire country, given the sheer number of the population.

93 *See* for an elaboration Spier, 2012, pp. 129 *et seq.* This is a very delicate and difficult issue to be pursued in more detail; that goes beyond the scope of this book.

94 *See* for elaboration on that point Spier, 2012, Chapters 10 and 12.

95 That is a huge topic in its own right. Legal systems greatly diverge on this point.

96 That view is in line with the OECD Guidelines for Multinational Enterprises, 2011, pp. 16, 29 and 30.

97 This stance is an elaboration of the point made in Spier, 2012, pp. 148 and 150.

emits considerably more than most of its competitors. In such a scenario, litigation may serve as an example for other small enterprises.

1.3.12 Injunctive Relief against Multinational Enterprises?

Multinational enterprises raise additional questions. Can a parent company be held legally responsible for the GHG emissions of the whole group (all its subsidiaries)? And, if so, can courts issue injunctive relief against a parent company to curb the GHG emissions of a subsidiary? Or should every single subsidiary be examined on its own emissions?

This question is largely a matter of the applicable (company) law.[98] That said, there is an **emerging trend that parent companies are under an obligation to ensure that their subsidiaries comply with legal obligations, particularly so in the realm of human rights;** all the more so if they know or ought to know that their subsidiaries probably violate their obligations.[99] The Ruggie Principles and the OECD Guidelines for Multi-National Enterprises may serve as an example. This, however, goes beyond the scope of this book. For now, it suffices that the issue is also important in relation to causation. If each subsidiary would have to be treated as a separate emitter, its contribution to the global GHG emissions would be exponentially low.

1.3.13 Various Emission Scenarios

This ultimately brings us to the hard core of causation. With the caveats mentioned before, I will confine myself to medium- and large-scale enterprises. As to countries, for practical purposes we can ignore the very poor and – population-wise – small countries such as the Vatican, and San Marino, given that their contribution is marginal. Besides, it is not overly likely that the courts of the poorest countries are willing to urge their governments to reduce their already limited GHG emissions, all the more so as this would largely go at the expense of the most vulnerable people. That leaves untouched, of course, that litigation targeted at major and hugely emitting *enterprises* in those countries may stand a chance; this goes particularly for subsidiaries of multinational enterprises with a seat in a 'developed' country.[100]

With the caveats just mentioned in mind, it may be useful to distinguish between five emission scenarios. In every scenario we assume that the State or enterprise violates its obligations to curb its GHG emissions to the extent legally required:

1. the emitter's emissions are very, very small; seen from a global perspective, less than 0.1 pro mille of the global GHG emissions;

98 *See* for an in-depth analysis L.F.H. Enneking, *Foreign Direct Liability and Beyond*, 2012.

99 *See inter alia*, M.T. Kamminga & S. Zia-Zarifi (Eds.), *Liability of Multinational Corporations under International Law*, 2000.

100 *See* for elaboration Spier, 2012, pp. 148 and 149.

2. the emitter's emissions are very small, say, more than 0.1 pro mille, but less than 0.1% of the global emissions;
3. the emitter's emissions are in the range of 0.1% and 1% of the global emissions;
4. the emitter's emissions are in the order of 1%-5% of the global emissions;[101]
5. the emitter's emissions exceed 5% of the global emissions.[102]

For the purpose of classifying a specific enterprise or State in one of the categories just mentioned, one has to answer the question of how to count the percentage. More specifically, should one count the *overall* GHG emissions of the respective player or only the *wrongful* part thereof? For most players, this hardly matters; their contributions remain very small, anyway. For States and enterprises in one of the categories 3, 4 or 5, it may have legal implications. The smaller the relevant contributions are, the more difficult it will be to obtain injunctive relief.

In my view the *overall* GHG emissions count in this respect. They serve as the factual and legal basis for determining how much a specific defendant must curb its emissions, given that the overall emissions do cause the problem. The wrongfulness issue remains important. After all, if the total emissions of a State or an enterprise do not exceed the limits allowed by law, injunctive relief must founder.

The argument that enterprises 'contribute' to climate change by their own GHG emissions is undisputedly true seen from a factual angle. One could, however, argue that the (ultimate) consumers are the real 'cause' of the emissions.[103] In the latter scenario, one could argue that:

(a) the 'ultimate' consumer 'substitutes' the manufacturer or
(b) the emissions of the 'ultimate' consumer when using the product have to be added to those of the manufacturer.

Both arguments (a and b) have their merits, but they make the attribution of GHG emissions unnecessarily complex. There may be instances where it is not unfair to add the emissions of a specific consumer to a specific 'manufacturer.' However, this strategy is not fair in each and every case. It *may* be appealing if a fast food chain seduces parents to take their young children to its unhealthy restaurants every day, and even more so if it adds some addicting substances to the food. But in many other instances, this approach is by no means self-explanatory. The same example also illustrates that alternative (a) above is not necessarily fair either. I am inclined to believe that both submissions should be ignored as they make the 'fight' against climate change by legal means too complicated. A lot would be gained if there would be a legal basis for submission (b). I wonder, however, whether the law as it stands (already) provides such a basis. In this respect, one should bear in mind that at the end of the day the consumer

101 By 2005, only eight countries (China, the US, the Russian Federation, India, Japan, Brazil, Germany and Canada) emitted 2% or more of the global aggregate; CRS report, p. 4 and Appendix B.

102 By 2005 only China, the US and the Russian Federation belonged to this group; CSR report, Appendix B.

103 The point is made, *inter alios*, by Mank, 2007, p. 205.

(the people at large) will have to settle the bill anyway, either because he will be confronted with the consequences of climate change or because the price of the products may (and is likely to) be increased if the manufacturing costs rise.

1.3.14 *Very Marginal Contributions (Less Than 0.1 Pro Mille of Global GHG Emissions)*

Most potential defendants belong to the first category (below 0.1 pro mille of the global GHG emissions). In those instances, plaintiffs seeking injunctive relief will find it difficult to persuade a court to award that relief. Causation to such a negligible extent is often considered insufficient,[104] although not universally.[105] The former, however, is a rather unsatisfactory finding. For practical purposes it would mean that a major part of the problem cannot

104 According to Art. 3:106 Principles of European Tort Law, 'minimal causation' has to be disregarded. Yet, that principle is written with pecuniary redress in mind (although that is not mentioned explicitly). *See* in more detail *Text and Commentary* (Spier), p. 59; Winiger *et al.* (Eds.), 2007, pp. 531 *et seq.* In his comparative analysis, Bernhard Koch observes that most reporters assume that their respective courts would deny causation (and therefore liability) in such cases altogether (p. 543); Koch points out that case law about this issue is scarce throughout Europe. The topic does not seem to be a hot topic on the agenda of academic writers in most jurisdictions (p. 542). Kysar, 2011, pp. 35 *et seq.*, footnote 172 and – about the related issue of joint and several liability for indivisible injury – footnote 173, also for further references. M. Loubser & R. Midgley (Eds.), *The Law of Delict in South Africa*, Oxford University Press, 2009, p. 75; C. van Dijk 'Civil Liability for Global Warming in the Netherlands', *in* M. Faure & M. Peeters (Eds.), *Climate Change Liability*, Cheltenham; Gloucester, Edward Elgar Publishing, 2011, p. 220; F, Trindade *et al.*, *The Law of Torts in Australia* (4th edn.), Oxford University Press, 2007, p. 541; A.M. Linden & B. Feldhusen, *Canadian Tort Law* (8th edn), Oxford University Press, 2006, p. 116, although it suffices that part of the injury is caused p. 120. Engel discusses a few nineteenth century public nuisance cases in which courts granted injunctive relief despite a very marginal contribution of the defendant; Engel, 2011, pp. 202 and 203. He advocates that view for climate change, be it in a somewhat narrow setting (*i.e.* "the proffer of third-party offset credits"; *see* p. 204 and for elaboration of his concept pp. 191 *et seq*). *See also* C. Sappideen & P. Vines (Eds.), *Feming's The Law of Torts*, Lawbook Co, 2011, p. 238. *See also* the rather critical judgment of the Federal Court of Australia in Wildlife Preservation Society of *Queensland Proserpine/Whitsunday Branch Inc v. Minister of the Environment & Heritage & Ors* [2006] FCA 736 (15 June 2006). Several leading authors seem to ignore these issues when discussing causation. Thus *e.g.* Grossman, 2003, pp. 22 *et seq.*; M.G. Faure & A. Nollkaemper, 'International Liability as an Instrument to Prevent and Compensate for Climate Change', *Stanford Environmental Law Journal*, Vol. 26a, 2007, at pp. 157 *et seq.* 2007, pp. 157 *et seq.*; Farber, 2008, pp. 401 *et seq.*; J. Smith & D. Shearman, *Climate Change Litigation*, Presidian, 2006. pp. 111 and 112; W. Braams *et al.*, *Klimaat en recht*, Deventer, Kluwer, 2010, pp. 12 and 13 in relation to "kleine vervuilers op wereldschaal" (small polluters on a world scale). According to Akkermans, there is no common ground for the submission that there should be a certain minimum contribution: A.J. Akkermans, *Proportionele aansprakelijkheid bij onzeker causaal verband*, Deventer, Tjeenk Willink, 1997, pp. 244 *et seq.*; it follows from his impressive comparative overview – in the context of loss of a chance – that a percentage below, say, 10% will rarely suffice (pp. 123 *et seq.*). According to Moréteau & Pellerin-Rugliano, French law rejects the notion of minimal causation in respect of requirements for liability, but does make use of it when it comes to determining compensation: *in* Winiger *et al.*, 2007, p. 533 (9/12). The Supreme Court of Canada requires reasonable foreseeability; *Mustapha v. Culligan of Canada*, 2008 SCC 27, *supra* note 13; it rightly observes that any harm that has actually occurred is "possible," so possibility does not suffice.

105 According to Giesen, 0.1% may suffice for damages; I. Giesen, 'Proportionaliteit in de klimaatdiscussie', *Nederlands Tijdschrift voor Burgerlijk Recht* 2012/51. His argument is based on the colossal aggregate losses. Vansweevelt & Weyts take the view that – in the loss of a chance context – there is no minimum threshold: T. Vansweevelt & B. Weyts, *Handboek Buitencontractueel Aansprakelijkheidsrecht*, Intersentia, 2009, no. 1022.

be solved, particularly in countries where injunctive relief against a State does not stand a chance. The latter is exactly the reason why the Draft of a new Austrian Liability Law is breaking new ground, in that marginal contributions suffice. The Commentary mentions losses caused by strikes, demonstrations or environmental damage caused by many tortfeasors. The Commentary points at the difficulty that the there is no measurable and noticeable difference in relation to the loss if we would ignore a specific tortfeasor. Given that the same applies to each single tortfeasor, the outcome would be that one faces a loss without any liable person. The draft Article 1294 para. 4 aims to overcome this problem to the extent legislation can ("soweit dies nach der Natur der Sache gesetzlich möglich ist").[106]

The above is not to say that there are no arguments in favour of a sufficient causal basis for injunctive relief, even against very small emitters. A Norwegian case may serve as an example. During World War II, a Norwegian political party sympathized with the German occupier. It was in government during the war and promoted the Nazi ideology. It also established armed forces to fight the Norwegian resistance and the allied forces. It caused a major loss of State and municipal money owing to corruption and maladministration. The defendant joined the party as a member in November 1940. After the war the State claimed compensation for the harm – estimated by the court to exceed 280 million Norwegian Crowns – done to the State. A majority of the Supreme Court found that the defendant was 'responsible' for the loss 'by joining' the party, although the Court admitted that the condicio sine qua non-requirement was not met. A causal relationship between the defendant's membership and each single wrongdoing was not required. The mere fact that he had joined and supported the party was enough. Not surprisingly, this judgment is rather controversial. It seems to stand alone.[107] Be that as it may, this judgment underscores that judges could be willing to break new ground in this area if they believe that a different outcome would be difficult – or impossible – to explain and justify. At least in the realm of climate change, I could not agree more.

A recent English case (*Rolls Royce Industrial Power (India) Ltd v. Cox*)[108] also seems to second the view that a very small contribution may suffice. Derek Cox was a welder. Between 1961 and 1985 he worked for a series of employers, one of them a predecessor of

The leading French textbook observes that in principle ("en principe") the French Supreme Court does not require (in my translation) that the judges dealing with the facts figure out whether the contribution in point is predominant, essential, or even that the loss was likely or even foreseeable ("Pas advantage la Cour de cassation n'exige en principe que les juges du fond relèvent le caractère 'preponderant' ou 'essentiel' de cette participation au préjudice, ni qu'ils constatent que le fait examiné rendait le dommage probable, vraisemblable ou même simplement prévisible", G. Viney & P. Jourdain, Traité de droit civil, *Les conditions de la responsabilité* (3rd edn), L.G.D.J., 2006, pp. 200 and 201, nr. 355); this, however, does not necessarily imply that there are no minimum thresholds. According to H. Luntz *et al.*, *Torts, Cases and Commentary*, LexisNexis; Butterworth Australia, 2009, a very small contribution may suffice (no. 4.2.12). According to Linden & Feldhusen, a chance of injury "in [1 of] 100 or even 1000" may suffice (*Canadian Tort Law* (8th edn), p. 131).

106 I. Griss, G. Kathrein & H. Koziol (Eds.), *Entwurf eines neuen österreichisches Schadenersatzrecht*, Springer, 2006, p. 44 (by Franz Bydlinski).

107 *See* B. Askeland, *in* Winiger *et al.*, 2007, pp. 534-536.

108 [2007] EWCA Civ 1189. *See*, more generally, M.A. Jones & A.M. Dugdale (Eds.), *Clerk & Lindsell on Torts* (20th edn), London, Sweet & Maxwell, 2010, nos. 2-39 *et seq.*

Rolls Royce. He died as a consequence of malignant mesothelioma. There was convincing evidence about the periods of employment in the service of the respective employers. Cox's widow sued Rolls Royce for damages. The Court of First Instance held that

> there is no reason to conclude that he [Cox] worked there [Rolls Royce's predecessor] just for a day or some other period that would be de minimis in relation to the exposure to asbestos. [. . .] On the balance of probabilities I conclude that it would be weeks at least [. . .].

In appeal, the latter finding was overturned as too speculative. Interestingly, Lord Justice Kay – seconded by Lord Justices Hooper and Tuckey – held

> **For the claim to succeed [UKP 100.000], the judge needed to be satisfied that** the extent and duration of the exposure had constituted a material increase in the risk to the Deceased of contracting mesothelioma. No specific measurement of the duration is necessary [. . .]. Exposure that falls within the *de minimis* formula would be insufficient. However, the type of contract work undertaken by [the predecessor of Rolls Royce] at power stations and the role of the Deceased in that work, coupled with his generic description of conditions in power stations at the time, undoubtedly justified the finding that this was not a *de minimis* case. Mr Limb [QC for Rolls Royce] frankly conceded that to work in such conditions for a week would not be de minimis.[109]

This judgment seems to suggest that very small contributions may suffice. Put differently, that the 'de minimis doctrine' is not necessarily a serious obstacle. This submission is reinforced, I think, by the strong language of Lord Justice Kay: "*undoubtedly* justified the finding that this was not a *de minimis* case" (first emphasis added).
A judgment issued by Justice Pain of the New South Wales Land and Environmental Court *may*[110] also foster the plaintiff's case if the emissions of the defendant are marginal:[111]

> The fact that there are many contributors globally does not mean the contribution from a single large source [. . .] in the context of NSW should be ignored in the environmental assessment process.

True, Justice Pain makes his observation in the context of an assessment process, but why should it be different in the realm of injunctive relief? The same goes for a judgment of the Victorian Civil and Administrative Tribunal:[112]

109 Jones & Dugdale, 2010, p. 21.
110 I have been unable to calculate the global contribution of the enterprise in point.
111 *Gray v. Minister for Planning* [2006] 152 LGERA 258 (Anvil Hill), quoted by Peel *et al.*, 2012, p. 271.
112 *Australian Conservation Foundation v. Minister for Planning* [2004] VCAT 2029 (29 October 2004).

> [. . .] the generation of greenhouse gases from a brown coal power station clearly has the potential to give rise to "significant" environmental effects.[113]

Although, admittedly, in a sense a side track, regard must be had to instances where courts easily find ways to circumvent the causation issue. An example, borrowed from Hart and Honoré's impressive book on Causation, may illustrate this point.[114] A driver who does not have the legally required driving licence causes an accident. Quite a few courts will find the driver liable for damages, despite the fact that it is very much in the clouds whether there is any causal relationship between the lack of a driving license and the accident.[115] This example shows that courts can reach an equitable outcome if they try hard enough. In itself, it is not a persuasive argument why a very small percentage can suffice, but it illustrates that it could if judges are willing to break new ground or to interpret the law in an innovative way.

Arguments borrowed from the law of 'wrongfulness' (negligence) may also foster the plaintiff's case. They will be discussed below in paragraph 1.4.4.

In the shorter term, we have to reckon with unfavourable judgments for the plaintiffs in litigation targeted at small emitters.[116] Do not misread me: the law as it stands is sufficiently flexible to assist victims, if courts would be willing to grant injunctive relief. Rules of minimal causation are (often?) not cast in stone, if cast at all. But bold judgments, overcoming the minimal causation obstacle, would come as a surprise for many and will be decried by many (lawyers) as the acme of judicial activism (which the latter do not mean as a compliment, be it that this reaction would not necessarily be justified). In a recent judgment, the US Supreme Court seems to suggest that it might be responsive to the argument that "ongoing constitutional violations [would] remain beyond redress,"[117] which seems to buttress the argument that at some stage courts will step in.

Engel has suggested that this obstacle can be removed by focusing on the last emission that was added. His argument goes as follows:

> Because the marginal damage curve slopes upwards from zero damages, each ton of greenhouse gas emitted imposes some measure of damages.[118] [. . .]

113 *Supra* note 43. This judgment is about considering the environmental effects resulting from the continued operation of a power station.

114 H.L.A. Hart & T. Honoré, *Causation in the Law* (2nd edn), Clarendon Press, 1990, p. 117. *Check met auteur: is dit de juiste ref.?*

115 *See* for more details Hart & Honoré, *Causation in the Law*, Clarendon Press, 1990, pp. 120 and 121. In some cases, it may be rather obvious that the lack of a licence is not the cause; *see e.g.* a Hong Kong case, discussed by Hart & Honoré, 1990, p. 119.

116 That is exactly the reason why Blomquist is very right in advocating that one should start with litigation against the major emitters: Blomquist, 2012, p. 1074.

117 *Los Angeles County, California v. Humphries et al.*, 562 U.S. (2010).

118 What is meant is probably 'damage'.

> First, the upward slope of the marginal damage curve means that the damages associated with the emission of any given ton of greenhouse gas will differ depending upon the concentration of greenhouse gases already in the atmosphere. [. . .] at a given point in time, the first ton of a greenhouse gas from anthropogenic sources added to the natural background concentrations of greenhouse gases, imposed the smallest amount of marginal damages, while the last such ton added imposed the greatest measure of measurable damage. We have no way of knowing, however, when the given emissions source emitted the greenhouse gas, and hence we cannot know whether the marginal damages attributable to a given defendant's emissions are relatively small or large.
> Thus the more accurate characterization of the damages attributable to a single source of greenhouse gas emissions is not that the damages are zero but that their size is indeterminate.[119]

His submissions are undoubtedly creative. They deserve warm support if they could serve as a vehicle to get at injunctive relief against small emitters. But courts unwilling to accept the view that marginal contributions suffice for injunctive relief may consider Engel's proposition rather artificial.
In brief, the chances of plaintiffs to seek for injunctive relief against defendants of the first category are not very favourable, though not hopeless if courts are willing to interpret the law in accordance with the urgent demands of our time. That was done on many other occasions in which the stakes were not (so) high.[120] Many courts have tried to do justice and to find a way to get at equitable results in a huge variety of cases.[121] Many of these judgments were applauded. It would be unsatisfactory if courts would not show similar courage in relation to the most urgent challenge of our time.

1.3.15 Loopholes to Reach a Desirable Outcome in Marginal Contributions Scenarios

If injunctive relief in relation to States or enterprises, whose contributions to the global emissions are marginal, would be likely to founder on the causation hurdle, one has to consider loopholes to obtain the desirable outcome. The submissions put forward below would need further elaboration and should be given further thought. But they may prove not to be unrealistic.

119 Engel, 2011, pp. 194-195.

120 *See*, in more detail, my contribution: 'The Rule of Law and Judicial Activism', *in* M. Faure & A. Van der Walt (Eds.), *Globalization and Private Law: The Way Forward*, Cheltenham, Northampton, Edward Elgar, 2010, pp. 426 *et seq.*

121 In relation to causation, Isachar Rosen-Zvi observes that "[g]enerally the Israeli Supreme Court tends to look for ways, sometimes very creatively, to compensate injured Parties when they believe such compensation is merited" (*in* Lord *et al.*, 2012, p. 291). This stance would be even more welcome in relation to injunctive relief.

There is an emerging line of thought that consumers should pay the cost of the impact on the environment of the products bought by them. The idea as such can hardly be contested as revolutionary. But for the reasons mentioned in paragraph 1.3.4, it will not be easy, or practical, to seek injunctive relief against private persons. Would it be a bridge too far to require from manufacturers that they raise the prices of their products to acquire proper equipment to reduce the GHG emissions caused by (the manufacturing of) their products?[122]

A less groundbreaking step would be injunctive relief against manufacturers of *unduly* energy eating and/or luxury products, unless produced (more or less) carbon neutral. That would be quite a step and the legal basis for this innovation would prima facie not be very solid.[123] But it might be worth a try before the 'right' court and on behalf of the 'right' plaintiff and against the 'right' (*i.e.* the 'wrong', in the sense just described) defendant. Blomquist, referring to Yukari Takamura, observes that there are persuasive Japanese precedents requiring industrial operations to shut down, "when there exists a serious threat to life and body."[124] If that is possible, it would be strange if a serious threat to millions of people would be struck out altogether.

Kysar points at automakers deliberately steering consumers towards high-emitting vehicles.[125] Would it be a bridge too far to grant an injunction to refrain from this practice? It would have at least some impact on the emissions of natural persons. A related and even more ambitious submission was made by Dubats. Striking a parallel with claims against the tobacco industry, she believes that claims against the hydrocarbon industry based on allegations of actively perpetuating misinformation about climate change might be capable of removing the causation obstacle.[126]

No doubt, defendants will argue that these submissions cannot serve to remove the causation obstacle. It cannot be denied that this is a very complicated issue and that there would be some truth in the defence that injunctive relief is not justified as long as the defendant's contribution to the threatened losses is very marginal. But there obviously is a difference between the common cases of, so to speak, 'average' States and enterprises, whose emissions are more or less in line with those of other States and enterprises in a similar position and enterprises (and States) engaging in an overload of GHG emissions and/or the production of unnecessarily energy-consuming products. The difference lies in the level of culpability, fault and/or wrongfulness (depending on the legal system). The latter carries and should carry weight and might serve as a vehicle to overcome the causation requirement.

---

122 Needless to say that enterprises may choose to make the investment without raising their prices.

123 *See* for a highly interesting discussion on this point Hunter & Salzman, 2007, pp. 1750 and 1751 (also quoted by Blomquist, 2012, p. 1058). *See* for other examples Blomquist, 2012, p. 1059.

124 *Ibid.*, p. 1075.

125 Kysar, 2011, p. 40.

126 E. Dubats, 'An Inconvenient Lie: Big Tobacco Was Put on Trial for Denying the Effects of Smoking; Is Climate Change Denial Off-Limits?', 7 *Northwestern Journal of Law & Social Policy*, Vol. 7, No. 2, 2012, p. 510.

So far, the discussion has tended to focus on injunctive relief sought by classes of potential victims or, if allowed by the applicable law, NGOs or other interest groups. They are the most obvious plaintiffs and, if well chosen, they are likely to appeal to most courts. This leaves untouched that other and at first sight less obvious plaintiffs could enter the scene. One could think of (groups of) enterprises, complying with their obligations to curb GHG emissions, who are affected by the distortion of the level playing field by competitors that did not (sufficiently) reduce their emissions. If the former (group of) enterprises would seek for injunctive relief against the latter, causation would not be an obstacle whatsoever. The financial impact of this distortion is obvious; it will often be well measurable and sufficiently significant. So even if the 'usual' classes of plaintiffs would be struck out by courts, enterprises belonging to the group of the willing could step in. Thus, they would act in their own interest and in the interest of society at large at the same time.
Conceptually, the same goes for classes of employees, banks and investors[127] of enterprises complying with their obligations who have reason to fear that the disruption of the level playing field will adversely affect them. Say, the carbon footprint of enterprise X is twice as high as the average of similar enterprises. Thanks to the huge GHG emissions, it can manufacture at much lower cost, so it drives competitors out of business. Pension fund P owns the shares of the impaired competitors. Why could they not seek injunctive relief against competitors that violate their obligations?[128] Once again, causation would not necessarily be a stumbling block if the opponents were well selected. Admittedly, this approach goes well beyond the beaten path, but it might well serve as a vehicle to circumvent causation, if necessary.
Be this as it may, I reiterate that the law in relation to injunctive relief is rather unsettled in relation to the requirement of causation. It is not beyond the realms of imagination that minimal causation would not be an insurmountable cliff.

1.3.16 Causation in the Range of 0.1%-1%

The third class of cases is less equivocal. In these cases it is certain that the contribution of a specific defendant lies in the order of 0.1%-1% of the global aggregate, *i.e.* a small, but still noticeable contribution; see also nr. 1.3.5 in fine.

127 If they have a majority interest in the enterprise in point, they could probably use their voting rights to achieve the result desired.

128 This submission presupposes that the defendants committed a (kind of a) tort towards (one or more of) the plaintiffs. Whether that is the case depends on the applicable law. Relativity (the German concept of *Schutznorm*) may come into play. Competition law might be called to aid, but other fields of the law might equally serve as a legal basis. More likely than not, courts could argue both ways (*i.e.* render plaintiff or defendant-friendly judgments), given that there are no precedents and that this still seems an area largely unexplored. In such instances, much depends on the preferences of the judges and their eagerness to break new ground and their courage to pave new avenues.

In the realm of damages, such a small contribution would probably not suffice in quite a few legal systems. Various common law countries require a "material contribution,"[129] although it seems a bit in the clouds what 'material' stands for; a small contribution in the range discussed here may not suffice. Leading authors, however, have advocated that small contributions to the global threat could, might or should be enough.[130] The Principles of European Tort Law suggest that a contribution of approximately 0.5% may suffice.[131]
In paragraph 1.3.5, I have already discussed a few cases that could support the victim's cause in the scenario in point. The following cases may serve as a further underpinning. First, an Estonian case, decided by the Supreme Court, seems to support the view that a contribution in the order in point may be sufficient. It was about a driver employed by various employers in the course of 31 years. In the course of his work he developed an occupational disease. The victim sued, among other defendants, an employer with whom the victim had worked for only eight or nine months. The defendant argued that this period was too short to have damaged the plaintiff's health. Unlike the courts of first and second instance, the Supreme Court "did not consider it correct to break down the compensation between the three defendants." Although the reported facts and the judgment, as reported, are not overly clear, it seems to support the view that marginal contributions may suffice.
One could equally strike a parallel with cases such as DES (a huge number of enterprises had manufactured a – as it turned out – defective drug; once the victims fell ill – years later – it could no longer be established, let alone proven, which enterprise had manufactured the drug used by [the mother of] a specific victim).[132] In the latter context a sheer *chance* of a causal link between a defective product of a specific manufacturer and a specific victim

129 *See, e.g.* Abbs *et al.*, *in* Lord *et al.*, 2012, pp. 94 *et seq.*; M. Dölle, D. Mahony & A. Smith, *in* Lord *et al.*, 2012, pp. 547/548. In the context of CERCLA (the Comprehensive Environmental Response) several courts have under certain circumstances imposed joint and several liability for cleaning up contaminated sites in relation to "concentric circles of third-, fourth- and fifth-party defendants, ultimately sometimes reaching into the hundreds": Gerrard, 2011.

130 Abbs *et al.*, *in* Lord *et al.*, 2012 , p. 97 (if I understand correctly); perhaps (but not very clear) Dölle *et al.*, *in* Lord *et al.*, 2012, p. 548; Farouk & Youssef, *in* Lord *et al.*, 2012, p. 265; Rosen-Zvi, *in* Lord *et al.*, 2012, pp. 292 and 293 (although not very clear); Takamura, *in* Lord *et al.*, 2012, pp. 229 and 230 (if I understand correctly); Glazewski & Collier, *in* Lord *et al.*, 2012, p. 340 (although not very clear); Goldberg & Lord, *in* Lord *et al.*, 2012, pp. 466-469. According to Shelton, 2005, p. 317, damage must have been caused in fact by the wrongdoer (but she does not explain what that means). Further down, she observes that "[i]n actions against the state, corrective justice theory may lead to presuming compensable harm when rights are violated, since the harm is inherently intangible and therefore impossible to prove". The latter clearly is a different issue, seen from a legal perspective, but one might also apply the "justice-argument" in the climate change setting; *see also Massachusetts v. Environmental Protection Agency* 549 U.S. 497 (2007). In J. Spier (Ed.), *Unification of Tort Law, Causation*, The Hague, Kluwer Law International, 2000, Suzanne Galand-Carval takes the view that 0.5% may be enough according to French law (p. 61). Helmut Koziol (the leading Austrian tort law expert) expresses the same view, speaking for himself (p. 20); with some hesitation the same goes for Italian law: F.D. Busnelli & G. Comandé, *in* J. Spier, 2000, p. 91.

131 Text and Commentary, Spier, p. 55, *supra* note 5.

132 *See* Kysar, 2011, pp. 37 and 39.

did suffice for pro rata, if not joint and several, liability in various jurisdictions.[133] If such a *chance* will support claims for damages, it would be difficult to justify why a *certainty* of threatening loss would be an insufficient basis to ward off the danger by means of injunctive relief, even if the contribution of each defendant is only in the range discussed here.

I readily admit that the submission that injunctive relief may stand a chance is not based on very stable ground. There is not (yet) a truly solid basis for such a submission; that is also the picture painted in the various contributions of the most valuable book on Climate Change Liability.[134]

On the other hand, the law as it stands in quite a few jurisdictions does not, let alone clearly, block injunctive relief. At the end of the day, it comes down to interpretation of the law of causation, arguably – but by no means necessarily – in a rather innovative way.[135] In this respect, the judgment of the US Court of Appeals in *Connecticut v. American Electric Power Company Inc.* may serve as a promising example. Next to the issues dealt with before, the Court addresses the defence that the defendant's emissions that "allegedly account for 2.5%[136] of man-made carbon dioxide emissions" are, in essence, too insignificant to cause future injuries, particularly since only "the collective effect of worldwide emissions allegedly causes injury." The Court labels this defence as "a variation on their argument that a polluter who 'contributes to' pollution does not allege causation," an argument the court has addressed and dismissed.

This issue, too, is addressed in the article by Kysar earlier quoted. In this respect also, his view is highly interesting. In the context of duty/proximate cause he discusses the theoretical possibility of climate change tipping points.[137] In these scenarios, "individual emitters may have to live with the possibility that their contribution was the one that pushed the planet over the edge." In his view there would be a common law precedent for such an exercise: asbestos, though scientists were unsure whether the underlying biological mechanism for asbestos-related cancers is a threshold one. He concludes:

> Because we do know that catastrophic thresholds loom with respect to climate change, the imaginative exercise may be *even more appropriate.*
>
> Moral imagination also is required to conceive of responsibility in terms that extend beyond the presentist and individualistic orientation of classical

133 *See* for a comparative overview Winiger *et al.*, 2007, pp. 448 *et seq.*; *see also* the respective country reports in Spier, 2000.

134 Lord *et al.*, 2012.

135 I endorse the view of Helmut Koziol (H. Koziol, *Basic Questions of Tort Law from a Germanic Perspective*, Jan Sramek Verlag, 2012, p. 272, no. 7/7) that it comes down to value judgments.

136 *I.e.* of five defendants together. We may fairly assume that each of them falls in the category discussed here (between 0.1 and 1% of the global emissions).

137 *See* about this issue also Spier, 2012, pp. 97 and 98.

> liberalism. [. . .] A different approach is to try to build new frameworks for the analysis of justice, ones that are commensurate with the temporal and geographic scale of the problem. Tort law is unlikely to be the venue within which such a project unfolds, but *it may not be immune from the project's influence*[138] (emphasis added).

Further down, he (rightly) observes that

> notwithstanding modern expansions, the concepts of duty of care and proximate cause in *tort law remain substantially animated by a classical liberal worldview* in which individuals are deemed free to pursue activities unless they impose harms upon identified victims that bear some geographic, market or other relational nexus to the actor. [. . .]
>
> Law in the twenty-first century will be preoccupied with the problem of instilling responsibility within complex networks. The challenge for tort law will be particularly acute [. . .][139]
>
> [. . .] climate change litigation may cause courts to *renew their romance with supra-individual conceptions* [. . .]as the mismatch between the atomistic account and our governance needs will become starkly apparent[140] (emphasis added).

Strong, stimulating and well-considered language. But it raises a conceptual question. First, was there a (solid) legal basis for the "classical worldview" that Kysar – and I – would be keen to overturn? Or was it just the court's appetite, whether based on the (alleged) demands or insights of those days or not? In the latter case – which seems more realistic than the opposing view – courts could adapt their reasoning and judgments to the demands of their own time without manipulating, let alone undermining, the law. Kysar's submissions seem to underscore this view. Courts could "renew" their romance with concepts that have been applied in various contexts. Why could these not be applied out in the realm of climate change (litigation)?
Turning to causation, Kysar observes that the limitations of the present case law

> are difficult to understand from an internal perspective on tort law. As commentators have noted, the *instrumental and fairness argument* in favour of market share liability *need not to be limited* to the kind of generic physical fungibility posed by DES. Other conceptions of fungibility – such as the generic lack of knowledge of any risk of HIV-AIDS among blood and plasma suppliers prior to the Center for Disease Control's warnings in 1983 – could just as easily

138 Kysar, 2011, pp. 51-52; footnotes omitted.
139 *Ibid.*, p. 54.
140 *Ibid.*, p. 55.

> support the market share theory. Likewise, the loss of a chance doctrine, at bottom, rests on *judicial recognition that certain increased risks can be sufficiently well-characterized to constitute a harm*, and that certain actors should be made liable for such harms when they stand in relations of heightened responsibility vis-à-vis their charges, even if the percentage change in the plaintiff's outlook does not clear a more-likely-than-not hurdle. These reasons need not to be limited to the medical malpractice context [. . .] More generally, increasing sophistication in the scientific understanding of risk counsels a shift from classical to probabilistic causation principles as a guiding template for analysis. Rather than but-for causation, *courts might take what Guido Calabresi calls "causal linkage" as the basic touchstone for analysis.* **At least, such a conception might be accepted in the public nuisance context when governmental plain-**tiffs allege an elevated risk of harm over a time series of events or within a population of individuals or resources. More ambitiously, courts might fashion a system of proportionate recovery by individuals for mere risk exposure, as commentators have long advocated[141] (emphasis added).

Unlike Doug Kysar, I do not advocate damages, not even on a proportional basis, for a number of reasons mentioned in paragraph 1.1.3. After all, and despite the fact that this stance is very difficult to explain to and justify towards victims, compensation is unmanageable and unbearable if we let matters proceed unchanged.[142] But *if* compensation were allowable, it would be difficult to understand and impossible to explain to the public at large (*i.e.* present and future victims) why injunctive relief that aims to *prevent* colossal harm, and by the same token human tragedies, all over the globe would not be an option. Once again, I basically agree with Kysar that the concepts mentioned in the quotation are based on legal interpretation of the law. For many – traditional – lawyers it will be quite a step to expand the same reasoning to climate change. But it can be done without overstretching well-established legal concepts. That said, I do appreciate that judicial courage will be needed. Courage shown by courts all over the world in many instances is of lesser importance.[143]

There are other arguments too to further the plaintiff's case. Injunctive relief is largely a matter of judicial discretion.[144] Although not universally accepted, injunctive relief may

---

141 *Ibid.*, pp. 63 and 64; footnotes deleted.

142 *See* for elaboration Spier, 2012, pp. 181 *et seq.*

143 *See* for examples J. Spier, 'The Rule of Law and Judicial Activism: Obstacles for Shaping the Law to Meet the Demands of a Civilized Society, Particularly in Relation to Climate Change', *in* Faure & Van der Walt, 2010, pp. 426 *et seq.*

144 *See*, among many other sources, US District Court Eastern District of New York in the Agent Orange case (*The Vietnam Association for victims of agent orange/dioxin et al. v. The Dow Chemical Company et al.*, also for further references); *Clerk & Lindsell on Torts*, 2010, nrs. 29-02 (and 29-4), adding that this does not mean that courts have an "arbitrary discretion"; Rogers, W.V.H., *Winfield & Jolowicz on Tort* (18th edn). Sweet & Maxwell, 2010, pp. 22-48. According to Markesini & Deakin (S. Deakin, A. Johnston & B. Markesinis, *Markesinis and Deakin's Tort Law* (6th edn). Oxford, Clarendon Press, 2008), policy matters cannot be

be *refused* if contrary to the public interest.[145] I wonder whether one could juxtapose the argument, in that the public interest might speak *in favour* of issuing injunctive relief, even if – perhaps – not all legal conditions are fully met.[146] There are three lines of thought that could be followed to get there:

1. judicial discretion absorbs causation. Put differently, issues such as minimal causation can (only) be weighed under the umbrella of judicial discretion, but they are not a decisive factor;
2. it would be unsatisfactory if the public interest could only play a role to *dismiss* injunctive relief. All the more so as the public interest in the latter kind of cases would often be relatively modest, compared with the high stakes in the arena of climate change;
3. in various fields – environmental issues included – class actions (or actiones populares) are gaining ground. In many instances, they aim only to serve the interests of a smaller group of people. But there is no convincing reason why they could not be conducted on behalf of a significant number of people, the inhabitants of an entire country or even future generations.[147] It would be against the spirit and the aim of this kind of actions if they were destined to be unsuccessful on merely doctrinal grounds in a field where action is required most.

True, this is an untraditional way of thinking. But why would it be more ground- breaking than the concept that the public interest could block injunctive relief?
Under English law, substantial harm is not required if

> the very fact that no harm is done is a reason for rather than against the granting of an injunction, for if there is no damage done the damage recovered in the action will be nominal and if injunction is refused the result will be no more nor less than a license to continue the tort [...] in return for a nominal payment.[148]

This argument carries weight in the field of climate change too.

---

evaluated properly "as long as we do not look beyond the formal conceptualism of the law – something that British judges and academics seem adverse to doing" (p. 24). More generally speaking, judges have quite some lee room in the realm of causation; *see e.g.* C. van Dam, *European Tort Law*, New York, Oxford University Press, 2006, no. 1111.

145 *See e.g. Weinberger v. Romero-Barcelo*, 456 U.S. 305, 312 [1982]. *See also* G.L.H. Fridman, *The Law of Torts in Canada* (2nd edn), Carswell, 2002, p. 205, referring to *Bottom v. Ontario Leaf Tobacco Ltd.* [1935] O.R. 205 (C.A.). This view is by no means generally accepted. Under English law, opinions are divided on the issue whether or not social and economic effects on third parties should be ignored: *Clerk & Lindsell on Torts*, 2010, pp. 29-09; *see also*, Sappideen & Vines, 2011, nr. 21.270, p. 522.

146 In *Crystal M. Rogers v. Commonwealth of Pennsylvania and Pennsylvania State Police*, the US District Court for the Eastern District of Pennsylvania makes the point that "the public interest is served if the plaintiff can file a complaint," but does not (need to) draw any conclusion from this finding.

147 *See* about the latter issue, Spier, 2012, pp. 55. *et seq.*

148 *Clerk & Lindsell on Torts*, 2010, p. 29-07, quoting *Woollerton and Wilson Ltd v. WH Smith (Eziot) Ltd.* [1987] 1 W.L.R. 853 at p. 860 per Balcome L.J.; the authors suggest that the better view is "that triviality of harm does preclude a prohibitory injunction" (29-07); *Winfield & Jolowicz on Tort*, 2006, pp. 22-48. Canadian law seems more demanding for the plaintiff: Fridman, 2002, p. 207.

Again a different approach to overcome the minimal causation hurdle is to borrow an argument from the requirement that injunctive relief may be *dismissed* if there is no irreparable harm, *i.e.* if the damage, if materialized, can be *compensated*.[149] Even if one were deaf to the moral argument that it is unsustainable that global devastation would have to materialize first before legal action can come into play, one should bear in mind that compensation of all future victims of climate change is simply impossible, because it would be far too costly. Seen from this angle, it does not come as a surprise that the US Supreme Court, following the line that injunctive relief is an equitable remedy and not a remedy "that issues as of course," takes the view that "[a]n injunction should issue only where intervention of a court of equity is essential in order effectually to protect property rights against injuries otherwise irremediable." It subsequently holds:

> Where plaintiff and defendant present competing claims of injury, the traditional function of equity has been to arrive at a "nice adjustment and reconciliation" between the competing claims. In such cases, the court "balances the conveniences of the parties and possible injuries to them according as they may be affected by the granting or withholding of the injunction." The essence of equity jurisdiction has been the power of the Chancellor to do equity and to mould each decree to the necessities of the particular case. Flexibility, rather than rigidity, has distinguished it. "In exercising their sound discretion, courts of equity should pay particular regard for the public consequences in employing the extraordinary remedy of injunction." [. . .] Moreover, the comprehensiveness of this equitable jurisdiction is not to be denied or limited in the absence of a clear and valid legislative command.[150]

If we would apply the test just cited, it seems to follow that injunctive relief might well stand a favourable chance. Balancing the respective interests can hardly lead to the conclusion that injunctive relief should be dismissed if the plaintiffs submit the right case. Put differently, the plaintiffs' interests should be convincing, *i.e.* the alleged threat of injury should be realistic and credible. That, I believe, makes the case of, say, the people of Bangladesh stronger than a case submitted by a few people who fear that, say, polar bears in their own region may disappear.[151] All this said, I understand perfectly well that injunctive relief is quite a step in relation to small emitters, also if the individual emissions are merely in the range discussed here

149 *See e.g.* the US District Court for the Eastern District of Pennsylvania in *Crystal M. Rogers v. Commonwealth of Pennsylvania et al.*, quoting Sampson, 415 U.S. at pp. 91-92. It obviously meant *Sampson v. Murray*, 415 U.S. 61 [1974].

150 *Weinberger v. Romero-Barcelo*, 456 U.S. 305 [1982]; references omitted. For good order's sake, the Court also observed that "The great principles of equity, securing complete justice, should not be yielded to light interferences, or doubtful construction."

151 To avoid misunderstanding, I have sympathy for the latter case, but if one tries to convince judges to deliver groundbreaking judgments, the case at hand should be as strong as possible. Speaking for myself, I find the case of Bangladesh stronger than the polar bear case.

(*i.e.* modest but well measurable). I do admit that the traditional view is not overly supportive for the submission that such a relief stands a chance. At the end of the day, lawyers – and particularly judges – carry a very heavy burden. Adhering to the traditional view would mean that they become co-responsible for global devastation. Leaving aside whether they should accept that fate if the law as it stands would not offer *any* manoeuvring room, the law in this area *is* sufficiently open and flexible to arrive at the morally imperative judgments. One should not easily accept that the law can only come into play in relation to – comparatively speaking – nitty-gritty cases. In this type of cases, lawyers should be willing to overstep traditional legal boundaries instead of being caught in narrow and often doctrinal boxes.
Goldberg and Lord hit the mark, observing that the argument

> that something actionable if done by one person is not actionable if done by many persons where each person only undertakes a small part of the action in question may – in their subtle English understatement – "be unattractive."[152]

Theoretically speaking, one could perhaps imagine another avenue, *i.e.* a kind of "specific performance" – either on the basis of the respective obligations of States and enterprises or on the basis of pledges many of them have made on various occasions.[153] Seen from a legal angle, such a position would have the advantage that causation (and quite a few other potential obstacles) would not be a hurdle to overcome. But, so far, the specific performance doctrine has largely focused on contractual relations. It is arguably a concept that deserves further exploration if injunctive relief would otherwise prove unavailable.

1.3.17 Very Small Contributions Ranging between 0.1 Pro Mille and 0.1% of Global Emissions

It follows from the above that the chances for injunctive victory are better in the third than in the first scenario. The range in point lies in between. Depending on the development of the law and the courage of courts, causation in this range may, or may not, turn out to be a major stumbling block.

152 Lord *et al.*, 2012, p. 469.

153 That is quite an issue in its own right; *see* Spier, 2012, p. 234. If this concept would work, it would only come into play in relation to national states as enterprises rarely have made *specific* pledges. Part of the problem will be that most pledges are rather vague; others are quite bold and promising, but it is open to debate whether most were meant to go beyond lip service to the common good. *See* about the enforceability and the status of these pledges (particularly in the context of the Millennium Development Goals): UNDP, *Human Rights and the Millennium Development Goals, Making the Link*, particularly pp. 13 and 16 and P. Alston, 'A Human Rights Perspective on the Millennium Development Goals', paper prepared as a conference contribution to the work of the Millennium Project Task Force on Poverty and Economic Development, available at <www2.ohchr.org/english/issues/millennium-development/docs/alston.doc>, particularly nrs. 35-42; after discussion of the respective views, he arrives at the conclusion "that it can plausibly be claimed that at least some of the MDGs reflect norms of customary international law" (nr. 42).

1.3.18 Contributions in the Range of 1%-5%

If the emissions of a specific defendant would be in the range of 1%-5%, the chances of getting injunctive relief granted are obviously (much) more favourable than in the scenarios discussed so far. If the lower emissions would suffice for injunctive relief, the same goes a fortiori for the range discussed here. The European Principles of Tort Law clearly suggest that a 1% contribution suffices.[154]
In an old US case against a mining company, the court held

> One drop of poison in a person's cup may have no injurious effect. But when a dozen, or twenty, or fifty,[155] each put in a drop, fatal results may follow. It **would not do to say that neither was to be held responsible.**[156]

If one would apply this approach, 2% would suffice for damages and, no doubt, a fortiori for injunctive relief. Why should the same not go for, say, 1%?
In this respect, the 'loss of chance' doctrine may shed some light on the realistic chances of injunctive relief. According to this doctrine, damages may be awarded in correlation with the chance (likelihood) that the alleged damage was brought about by the defendant's wrongful conduct. The doctrine is by no means common ground. It is, *e.g.* (largely) rejected under Australian[157] law. Opinions are divided in the United States. This concept, however, gained ground in many European countries[158] and is also adopted in the European Principles of Tort Law.[159] Several Supreme Courts in the United States have equally adopted the concept.[160] Opinions diverge as to the threshold.[161] But in quite a few

154 Text and Commentary, Spier, p. 55, *supra* note 4 . Kaminskaite-Salters takes the view that causation in this range does not suffice (for damages) under English law as it stands, Kaminskaite-Salters, 2010, pp. 181 and 182.

155 *I.e.* 2%.

156 *Woodyear v. Schaeffer*, 57 Md 1 (Md 1881) at pp. 9 *et seq.*, quoted by Verheyen, 2005, p. 255. *See* for another example (about approximately 4%) *Warren v. Parkhurst*, 92 N.Y.S. (N.Y. Supreme Court 1904), aff'd 93 N.Y.S. 1009 (1905), quoted by D.W. Robertson, 'Causation in the Third Torts Restatement: Three Arguable Mistakes', available at <http://tortsymposium.law.wfu.edu/papers/robertson.pdf>. Akkermans refers to *Sykes v. Ministry of Defence*, in which case 2%-3% did suffice: Akkermans, 1997, p. 247.

157 *Tabet v. Gett* [2010] HCA 12 (21 April 2010) in relation to medical malpractice. Strikingly, the doctrine does apply in commercial cases.

158 *See* for more details Koziol *in* Winiger *et al.*, 2007, p. 589. As to the Netherlands, it is settled case law that a chance suffices; *see e.g.* Hoge Raad (Supreme Court) 24 October 1997, *NJ* 1998/257 and Hoge Raad 16 February 2007, *NJ* 2007/256.

159 Art. 3:106 PETL.

160 *E.g.* Supreme Court of Louisiana, *Rachel Smith v. State of Louisiana*; Supreme Judicial Court of Massachusetts (*Matsuyama v. Birnbaum* of 23 July 2008); the Court discusses the doctrine at length, referring to a wealth of academic writings and case law; the court recognizes the doctrine as a theory of injury; it stresses that the application is restricted to medical malpractice. Other Supreme Courts have rejected the loss of a chance-doctrine; thus *e.g.* Supreme Court of Texas, *Kramer v. Lewisville Memorial Hospital*, 858 S.W. 2d 397 (Texas 1993).

161 *See* Koziol, *in* Winiger, 2007

countries, a chance in the range of 1%-5% may do.[162] If that holds true for claims for damages, it rather belabours the obvious that it should equally do so in relation to injunctive relief. Even more so as the stakes in the area of climate change are so high.

1.3.19 Contributions in Excess of 5%

It follows from the discussion of the previous category that the chances for injunctive relief are in theory quite favourable if the defendant's GHG emissions exceed 5% of the global aggregate. An old English case – in the realm of contract law – may serve as an example.[163] Chaplin, along with 6,000 others, entered a nationwide beauty contest and got to the final stage of only fifty contestants. The winners were given theatrical engagements for three years at UKP 5 per week. Hicks, in breach of his contract, prevented Chaplin from taking part in the final selection. The judge and the jury awarded her UKP 100 for the lost opportunity. That judgment was upheld in appeal. Lord Justice Fletcher Moulton observed that the injury was "very substantial." Those days, UKP 100 was quite something, but if compared with the stakes of climate change, it is trivial. So the same should a fortiori be accepted for climate change, I think.

As a matter of fact, plaintiffs run a fair chance of losing their case as the number of countries concerned is small. Only China, the US, the Russian Federation and India fall in this category.

It does not seem overly likely that Chinese or Russian courts will be prepared to grant injunctive relief against the respective States. The US position is less sure; much may depend on the composition of the US Supreme Court at the time of the delivery of the judgment. Litigation in India may stand a better chance, given the admirable courage of the Indian Supreme Court.[164] But the Indian GHG emissions per capita are still quite modest, so (seen from a global perspective) at the very best, the reductions to be ordered will be modest.

1.3.20 Can the Hurdle Be Taken by Suing Several Defendants before the Same Court?

Kaminskaite-Salters suggests that if a series of defendants would be sued before one and the same court, their cumulative contributions might count.[165] The idea is certainly appealing, but I wonder whether there is a legal basis for this submission. It would contribute to a fair solution in the field of climate change, but in other scenarios, it seems a rather dangerous approach. Put in extremis, one can hardly imagine that it would work to

162 The Supreme Court of Louisiana held that "any degree" is compensable (*Rachel Smith v. State of Louisiana*). According to Akkermans, 1% may be enough, if that percentage is sufficiently certain; Akkermans, 1997, p. 253.

163 *Chaplin v. Hicks* [1911-1913] All ER Rep. 224 per Fletcher Moulton LJ.

164 *See* Rajamani & Ghosh, *in* Lord *et al.*, 2012, pp. 176 and 177.

165 Kaminskaite-Salters, 2010, p. 81.

sue all American citizens and that their respective contributions should be added up, thus that the aggregate would count as the contributions of each single citizen, or, alternatively, that the minimal causation hurdle could be removed by the mere fact that more people are sued at the same time before the same court.
She also points at the possible option of alleging that defendants are acting in concert, but rightly adds that this will be extremely difficult to prove.[166] Besides, it would end up in joint and several liability – in my submission a very unbalanced outcome. It is less clear what this approach would mean in the context of injunctive relief. I do not think that there would be much support for the view that each and every defendant would have to assume the obligations to reduce the emissions of all the other defendants, which would mean that they would all be driven out of business.[167]

1.3.21 Conclusion

It follows from the above that causation may be a serious obstacle for injunctive relief, particularly so in relation to (most) enterprises and the vast majority of States. But not necessarily so. First, it could be argued that (minimal) causation is not a *requirement* for injunctive relief. Secondly, there are precedents supporting the view that this hurdle can be overcome.
Given that the law (of causation) is still rather unsettled in this field, it may prove useful to focus on a related issue: wrongfulness. In that context similar questions arise, *i.e.* does a minor chance that damage will occur or a very small contribution to a loss suffice for labelling an act or omissions 'wrongful'? Admittedly, the mere fact that the answer would be in the affirmative does not, let alone necessarily, mean that a similar loose causal relationship will or should suffice to achieve injunctive relief. But at least in the context of climate change it would be unsatisfactory if a wrongful act or omission could not be stopped.[168] After all, it would mean that there would be no remedy for wrongful situations, not even if they affect society at large and by the same token a great many million people, present and future. Traditional lawyers may not care about such a scenario, trained as they are to find their way in the subtleties and the twisting roads of the law. But such a position would not be easy to explain, let alone to justify, to ordinary people. In this context, it should be borne in mind that it is at least open to debate whether the idea that there would not be any legal remedy to prevent colossal damage for society at large, including the right to life for many people, is difficult to reconcile with the essential right of access to court; although I know that this right is often interpreted in a somewhat restrictive way.

166 Kaminskaite-Salters, 2010, pp. 182 and 183.

167 Not surprisingly, Kaminskaite-Salters acknowledges that joint and several liability "may have catastrophic consequences" (pp. 183 and 184). I assume that "may" should be read as "will."

168 In *Mustapha v. Culligan of Canada* the Supreme Court of Canada hints at the intertwining of both legal concepts: 2008 SCC 27, *supra* note 13.

1.4 The Standard of Care, Also in Relation to Minimal Contributions

1.4.1 Why a Focus on States Is the Most and at the Same Time the Least Promising Way Ahead

Even if we ignore, for a while, the diverging responsibilities and legal obligations of the respective *States*, the only feasible way to get at the reductions needed is to focus on States. Whatever the legal obligations of the respective elements of society ('ordinary' people, 'the government' and enterprises) may be, it is arguably of limited practical use to examine the liabilities of the 'ordinary' people and a major part of enterprises, given that it will be fraught with serious difficulties to enforce their obligations. This is not to suggest that 'ordinary people' and small enterprises do not have obligations in this field; they do. But national States enjoy at least some margin of appreciation to allocate parts of the burden to the respective players in their jurisdictions. Even if we could ignore the margin of appreciation, for example because a national State refrains from any allocation, enforcement is a rather theoretical vehicle if these actors do not willingly comply with their obligations, if not for other reasons because litigation is too costly and too burdensome.

This leaves untouched that it remains useful to indicate their obligations clearly. After all, individuals and small enterprises *may* feel obliged to meet their obligations if those obligations are sufficiently clear. In the short term it is not overly realistic to assume that many will change their lifestyle. We may take it for granted that many people already know quite well that their lifestyle is not sustainable. They may well care in the abstract, but so far they do not feel obliged to change course. Why would they change their attitude if they realized that they are legally bound to reduce their carbon footprint with a *specific* percentage?

Besides, it is rather difficult, if not outright tricky, to determine the obligations of 'ordinary' people, even if we limit our enquiry to citizens of 'developed' nations. The determination of their obligations gives rise to myriad questions, such as:

(a) is each citizen entitled to the same carbon footprint?
(b) if so, is a couple with – say – five children entitled to seven times the 'emissions' compared with a single person? If not, what then are their 'emission rights?'
(c) do people with small incomes have the same 'emission rights' as better paid people?
(d) do (small) children have the same rights and obligations as elderly and/or ill people?
(e) does it matter whether one lives in a dry, wet, cold or hot climate?

I am afraid that the law falls short in answering these questions at the level of individual citizens. These ultimately are political issues to be dealt with by politicians. That remains true even though politicians have fallen grievously short in answering these questions. It follows that it is of the utmost importance to extend our enquiry to countries. Only they can ensure that the aggregate obligations of *all* national 'players', be it enterprises, citizens or governmental institutions, will be met.

By the same token, it is of significant use to establish the obligations of the *respective national States*. That is already a major challenge in itself. It will be discussed below.

As an unfortunate matter of fact, (most) States will not solve the problem, let alone in the short term. It would require bold and immediate action of politicians. Only very few of these are truly willing to take an active stance in this debate, let alone taking appropriate action. Worse, it is unrealistic to assume that the obligations of the major emitters (China, the US, the Russian Federation and India) can be enforced by legal means.[169] States can only be sued before national courts.[170] In the end, the courts of two of these countries are not (sufficiently) independent; another superior court has shown utmost reluctance, adhering to the political argument line. The remaining Supreme Court belongs to the most activist in the world, but the GHG emissions of its country are relatively modest, particularly if counted per capita. So even if that Supreme Court would be willing to urge the State to change course, this will only have a small impact. According to Rajamani it is a matter of time "before climate-based rhetoric and arguments move from the periphery to the core of the mudicial process." She casts doubts whether this would be a task for the courts.[171]

This unfortunate state of affairs begs the question whether the countries with judges who understand the demands of our time could and should fill the gap, which would mean that their countries would have to assume the burden of the reductions not brought about by countries that fail to meet their obligations, while these obligations cannot be legally enforced. I will discuss that issue below.

The above does not mean that it is useless to submit a blueprint of the obligations of the respective States, even if they cannot be enforced by legal means. A credible picture could serve other purposes, such as an underpinning of the bargaining positions of the most vulnerable and most imperilled States. At some stage, it may also cause sufficient embarrassment to bring reluctant politicians to their senses. Thus, although the ultimate goal to come to grips with the looming threats of climate change can probably not be reached in due time, such a blueprint may well have a useful impact.

1.4.2 A Focus on Enterprises

A focus on (major) enterprises stands more favourable chances for quite a few reasons. First and foremost, it is not – or at least, less of – a political issue. So courts may feel less reluctant to step in. Secondly, enterprises cannot be sued before their national courts

169 Interestingly, two US Senators (Boxer and Sanders) have introduced climate change legislation that aims to reduce GHG emissions considerably; *see The Guardian* of 16 February 2013.

170 The ICJ hardly plays a role in relation to these countries, even if it would be willing to render bold and courageous judgments. In Europe and the Americas, Human Rights courts could – and hopefully will – play a major role, but the difficulty lies in the enforcement of their judgments.

171 Rajamani, 2013.

only.[172] More importantly, subsidiaries or branch offices can be sued before the courts in countries where they are based. This provides opportunities for plaintiffs in countries with more activist courts. Thirdly, many enterprises have assets abroad; they can be seized if litigation would prove necessary. Last, but not the least, a growing number of executives understand the change of the tide. They do care for the good cause, for substantive and/or public relation reasons. They are absolutely right, if not for other reasons because major investors are getting ever more worried about the devastation climate change is going to bring and the potential adverse impact of claims for damages on 'their' enterprises. They can and undoubtedly will put increasing pressure on enterprises to meet their legal obligations to avoid this kind of litigation.

All this said, a focus on enterprises cannot solve the entire problem. After all, they are only 'liable' for part of the global GHG emissions. One could possibly argue that they are responsible not only for their own excessive GHG emissions,[173] but equally for the adverse impact of their goods and services, if used by consumers;[174] see for an elaboration paragraphs 1.3.11 and 1.3.13. Even in that scenario, we still cannot achieve a sufficient level of reductions of the global GHG emissions, if for no other reason than that a major part of the goods and services is delivered by small(er) enterprises. They, too, have legal obligations. But enforcement is fraught with difficulties, as mentioned above. The shares of many of the small(er) enterprises are not held by major investment funds, so they cannot put any pressure on this category of enterprises.

It follows that a focus on enterprises can only solve part of the problem, albeit a not unimportant part.

1.4.3 Common but Differentiated Responsibilities

According to the (still?)[175] prevailing view, States have common but differentiated responsibilities (hereinafter cbdr). That is, their legal obligations to curb their GHG emissions diverge. The justification for this approach lies in the significantly diverging historical contributions and in the level of 'welfare'. Many so-called developing nations have hardly contributed to the emerging threats. Even these days, their contributions are relatively marginal and in some instances only modest, and even more so if counted per capita.

172 Whether they can or not depends on the national law of the court in point. Depending on the applicable law, parent companies that have a seat abroad may be 'immune' from litigation in other countries. Whether they are may depend on the question whether a court is willing to accept the argument that excessive GHG emissions (may) cause damage in its country. *See* para. 1.3.12 for the parent company issue.

173 A shorthand for emissions over and above the quantum legally allowed.

174 I am inclined to believe that this argument is invalid for manufacturers earlier in the chain, such as manufacturers of raw materials or components to be incorporated in other products. Otherwise, one and the same emission would count twice or even more often.

175 *See* about its fortunes L. Rajamani, 'The Changing Fortunes of Differential Treatment in the Evolution of International Environmental Law', *International Affairs*, Vol. 88, No. 3, 2012, pp. 605-623.

For our purposes, this also is a tremendously important topic. First, it is seen from the angle of fairness. It is crucial to strike a fair balance between the greatly diverging interests and needs of the respective states. But also because our goal is to come to grips with the looming threats of climate change. I have already mentioned that this means that the *global* GHG emissions must be reduced at a great pace and to a considerable extent. If a series of countries is not (yet) under an obligation to curb their emissions, or alternatively only to a limited extent, the other countries will have to assume their part of the burden. Otherwise, our attempt to come to grips with climate change is doomed to fail from the beginning. Seen from a moral and practical angle, that does not need further explanation. But is it also a legal obligation? I will address that question below.
Be that as it may, **assuming responsibilities of others presupposes** that one knows the **scope of the responsibilities of the vulnerable countries that do not, or only marginally,** have to contribute and the size of the reductions required from emerging economies that have to contribute less.
It is arguably open to debate whether the cbdr paradigm is still universally adopted. But even if it is, it is very uncertain what it really means. What are the obligations of the respective nations? The Annexes to the Kyoto Protocol give at least a hint of how to determine the respective contributions. But it remains a slippery slope.
Quite a few academics, research institutions and NGOs have submitted more or less concrete ideas on how to determine the respective obligations. Chapter 13 of IPPC's Climate Change 2007 report on Mitigation,[176] provides an excellent concise overview of a series of submissions.[177] The issue is discussed at some length in my Shaping the Law for Global Crises.[178]
I wonder whether it is very helpful to discuss this issue in the context of wrongfulness.[179] After all, it is very much open to debate whether private law can provide any credible underpinning of the respective obligations, departing from the cbdr maxim. First, it is a global concept, while there hardly is a generally accepted basis in this field. It would be unsatisfactory and not overly useful to attach very different meanings to the relevant concepts, depending on the *national* private law in point. Secondly, the answer depends, at least to some extent, on the *intentions* of those who have crafted the concept of cbdr and the interpretation of the international 'consensus'. The latter is a delicate and very complex issue in its own right. First, 'developed' and 'developing' nations do and did interpret the concept in a different way. Secondly, the ideas have progressed over time, and it may well be unclear to what effect.
With all these provisos, the common denominator of private law can arguably be of some help, as I will try to demonstrate below. At this stage it may suffice to observe that there seemingly is an emerging common ground that the following factors play a role in determining the obligations of the respective countries:

176 S. Gupta *et al.*, Contribution of Working Group III to the Fourth Assessment Report of the Intergovernmental Panel on Climate Change.
177 *See* IPCC, 2007, pp. 770 *et seq.*
178 *Ibid.*, pp. 117 *et seq.*
179 I mean: wrongfulness according to private law.

(a) the country's capabilities in terms of wealth and needs;
(b) historical contributions;
(c) dependence on fossil fuel and access to renewable energy;
(d) emissions per capita.[180]

'Developing' countries may be in a position to reduce GHG emissions without an adverse impact on the economy and by the same token on their most vulnerable people. The "New Vision for Agriculture" may serve as an example. It would imply a 20% increase in production, a 20% reduction of emissions and a 20% poverty reduction. However, it would require investments.[181] India could eliminate blackouts by 2014 and increase the value of the national output by 50% by 2017 through strong energy-efficiency standards.[182] If these and other 'findings' would be correct for 'developing' countries in general, they could arguably be required to reduce their GHG emissions, particularly so if the other countries (or international institutions such as the World Bank) would be willing to provide the necessary funds. Lord Stern expresses a firmer view: "developing countries must take on the lion's share of cuts in GHG emissions, because of the 'brutal arithmetic' of climate change." After all, even if "developed countries" would reduce their emissions to zero, "that would not be enough to halt runaway climate change – because emissions from rapidly industrializing economies are now so high." Thus he is "not pointing a finger at the developing world, just looking at what is necessary." New technology would be the answer.[183]
Stern probably has a point. But his submissions do not stand a favourable chance and – at least in the short(er) term – it would also be unfair, unless "developed countries" are willing to provide the funds and/or the technology to enable countries such as India (and arguably also China) to achieve the reductions necessary to attain the imperative total of reductions to avoid catastrophes worldwide. I am inclined to believe that there is a legal basis for this submission, as will be briefly discussed below.

1.4.4 *Minimal Contributions as an Obstacle for Wrongfulness?*

1.4.4.1 Introduction

In the previous paragraphs, I have discussed the question whether minor contributions suffice, seen from the perspective of *causation*.[184] I arrived at the conclusion that this issue largely is new ground that needs further exploration. If a condicio sine qua non-nexus

180 *See* in more detail, also for further references, Spier, 2012, pp. 129 *et seq.*
181 *See* N. Stern, 'How We Can Respond and Prosper: The New Energy-Industrial Revolution', Lionel Robbins Memorial Lecture Series, Lecture 2, 22 February 2012, p. 43.
182 Stern, 2012, p. 16.
183 *The Guardian*, 4 December 2012, Lord Stern: developing countries must make deeper emission cuts.
184 Luntz, Hambly, Burns, Dietrich and Foster observed that – in the realm of causation – a material increase of the *risk* may suffice, whereas a material contribution *to the injury* is not required; Luntz *et al.*, 2009, nos. 4.2.13 and 4.2.17.

between the emissions of a specific actor and (the threat of) global climate change would be a requirement for injunctive relief, claims *may*[185] founder on this obstacle.

If causation would not be an obstacle, either because it is not a requirement for injunctive relief, or because the obstacle could be overcome, we face a new challenge: wrongfulness. As such, it seems to belabour the obvious that exposing present and future generations to the most grievous harm ever is wrongful. But things are not that easy, I am afraid.

The difficulty lies, once again, in the small or even marginal – if measurable at all – contributions of most actors, particularly so the overwhelming part of enterprises.[186] One cannot take it for granted that exposing others to marginal harm, if counted per actor, is wrongful. I am inclined to believe that this topic also falls under the umbrella of the duty of care (wrongfulness), be it that it should be distinguished from the traditional questions. If national (tort) law would fall short to address marginal contributions of each single actor, other parts of the law could arguably be called upon. If, for example, a State or enterprise would violate one or more human rights, despite its minimal contribution to the global GHG emissions, this violation of the law would 'colour' the national requirement of tort law, I think. Or, to be more precise, if excessive GHG emissions by a specific actor would be a violation of human rights – as probably is the case – those emissions would also amount to a tort, apparently irrespective of the marginal contribution of the specific actor to the global problem. That would be a double catch. First, for practical purposes, human rights could be enforced by national courts, and secondly, national (tort) law could give a more *specific*, and detailed, meaning to the rather vague notions of general human rights. As things stand now, it is open to debate whether the respective human rights, such as the right to life and a series of social and economic rights, point to *sufficiently detailed obligations* of the respective countries (and enterprises), I think.

As to the enforcement of human rights (and, I tend to believe, of wrongfulness coloured by human rights), the Ruggie Principles point to a duty, vested in States, to protect against human rights abuses by "third parties," including businesses and State-owned enterprises.[187] That is quite helpful, as was possibly underestimated by politicians when they enthusiastically endorsed these principles.

1.4.4.2 'Precedents'

Back to tort law. There is some doctrine and case law about the following set of cases: damage that occurred due to the fact that remote chances materialized. Say, a drunken tramp is hit in the middle of the night in the middle of nowhere because he crossed a street in a small village at the very moment that a car coincidentally passed by. Or someone gets

185 May, as it is unclear whether causation is a requirement for injunctive relief; *see* para. 1.3.4.
186 *See* for a similar view Van Dijk, 2011, p. 216.
187 *See* in more detail Spier, 2012, pp. 38 and 39.

poisoned by the remains of a snail in a bottle of ginger beer.[188] Seen from a traditional (doctrinal) angle, those cases are very different from climate change. They are, among other issues, about remote *chances* that materialize. Climate change is about minor *contributions* that, together with the emissions by many others, cause or contribute to serious harm on a global scale. The latter scenario is comparable with pollution of a river by the entire industry which discharges its waste or pollutants into the river, whereas the contribution of each actor is marginal; or with the immense and very dangerous discharge of plastics in the sea by a great many countries and enterprises.[189] In the latter set of cases, doctrine and case law is still in its infancy. So if we are looking for useful 'parallels', we have to stick to settled case law that, in a sense, comes as close as possible to potential climate change cases, despite the obvious and not unimportant doctrinal differences.

In paragraphs 1.3.14 and 1.3.16, I have already discussed a few cases that seem to suggest that remote possibilities or minor chances may be enough for establishing a condicio sine qua non-nexus in a specific case. The same goes for wrongfulness, as will be demonstrated below. For our purposes, we have to delve into the basic requirements for wrongfulness, which also play an important role in determining the obligations of the respective players. The yardstick for the assessment whether a specific act or omission is (un)lawful or not is whether the actor meets what the standard of a 'reasonable person' (*bonus pater familias*) would/should have done in similar circumstances.[190] What can reasonably be required of such a person?[191] The European Principles on Tort Law elaborate on this topic as follows. It depends

> in particular, on the nature and the value of the protected interest involved, the dangerousness of the activity, the expertise to be expected of a person carrying it on, the foreseeability of the damage, the relationship between those involved, as well as the costs of precautionary or alternative methods.[192]

In California the policy of preventing future harm also carries weight.[193] The same probably goes in many other jurisdictions. Under Australian law, the social utility of the risky activity plays a role.[194]

188 *Donoghue v. Stephenson* [1932] AC 562.

189 *See* M. Allsopp *et al.*, 'Plastic Debris in the World's Oceans', available at <www.unep.org/.../docs/plastic_ocean_report.pdf>; J.B.G. Derraik, 'The Pollution of the Marine Environment by Plastic Debris: A Review', *Marine Pollution Bulletin*, Vol. 44, 2002, pp. 842-852.

190 *See e.g.* paras 291-293, Restatement Second on Torts and Art. 3:102 PEL Liab. Dam.

191 *See* in more detail Van Dam, 2006, pp. 189 *et seq.* Part of the text is borrowed from Spier, 2012, pp. 88 *et seq.*

192 Art. 4:102, para. 1. As just mentioned Art. 3:102 PEL Liab. Dam. is less explicit. According to the commentary, "several factors beyond conclusive enumeration" play a role. Courts have to assess what constitutes careful conduct in a given set of factors that may change over time; *see* Von Bar, 2009, pp. 585 and 586. I wonder whether judges really assess what would have been *proper* conduct. Instead, I think, they often go no further than labelling a specific act or omission as wrongful. Put differently, they often intuitively jump to conclusions.

193 *Rowland v. Christian*, referred to in V.E. Schwartz, K. Kelly, & D.F. Partlett, *Prosser, Wade and Schwartz' Torts, Cases and Materials,* Foundation Press, 2010, p. 147.

194 Trindade *et al.*, 2007, pp. 433 and 434.

A similar approach is adopted in, for instance, the US (the so-called Learned Hand formula),[195] England and Wales,[196] Canada,[197] New Zealand,[198] Australia,[199] the Netherlands[200] and South Africa.[201] The International Commission of Jurists seems to take a similar position.[202] Canadian Tort Law very much depends on foreseeability.[203] It "seeks to impose a result that is fair to both the plaintiffs and the defendants, and that is socially useful,"[204] thus leaving considerable leeway for courts.

The ICJ persuasively puts forward the view that in considering what a prudent company would have foreseen, the court will look at "objective evidence as to what kind of information was available to the company about the risk [. . .] from its own employees and consultants, the media and civil society." Besides, "a reasonable person in the company's shoes would have undertaken an inquiry as to the potential risks involved."[205]

Not surprisingly, the "Learned Hand formula" is gratefully adopted by proponents of and believers in law and economics.[206] I do not belong to the camp of the believers, but the theory could serve as an underpinning for the case of victims.

If we ignore the issue of "discounting," popular in the world of law and economics – a concept unfit for our purpose, given the human tragedies if we continue doing "business as usual"[207] – one cannot escape from the very strong impression that leaning backwards is much more detrimental to society than taking bold action, even if one confined oneself to economic interests. The Stern report is telling in this context. It may be true, as Stern has readily admitted recently, that his report is outdated and that it is based on overly optimistic scenarios, things having become even worse. It may well be possible – Stern is a bit ambiguous on this point – that the cost of taking action will exceed the level of 2% of 3% of GDP, but the cost of leaning backwards is exponentially higher. In Stern's 2006 report the costs of continuing business as usual (the adverse impact on the economy) were estimated at 20% of the global GDP in a worst-case scenario. Stern now believes that they would be

195 *US v. Carroll Towing* [1947] 159 F(2d) 169, 173; *see* in more detail D.B. Dobbs, *The Law of Torts*, West Group, 2000, §145. *See*, also about the draft Restatement Third, K.W. Simons, 'The Hand Formula in the Draft (Third) of Torts: Encompassing Fairness as Well as Efficiency Values', *Vanderbilt Law Review*, Vol. 54, No. 3, 2001, p. 901. *See* for a further elaboration also *Prosser, Wade and Schwartz' Torts*, 2010, pp. 147 *et seq.*

196 *Morris v. West Hartlepoole Co. Ltd.* [1956] AC 552, 574 per Lord Reid; the judgment adds the use to society.

197 Linden & Feldhusen, 2006, p. 130.

198 W.C. Hodge *et al.*, *Torts in New Zealand, Cases and Materials* (2nd edn), Oxford University Press, 1997, pp. 212 *et seq.*

199 Trindade & Cane (3rd edn), pp. 341 *et seq.* and Trindade *et al.* (4th edn), pp. 428 *et seq.*

200 Settled case law since Hoge Raad 5 November 1965, *NJ* 1966/136. *See* about this case law in the context of climate change, Van Dijk, 2011, pp. 214 *et seq.*

201 J. Neethling, J.M. Potgieter & P.J. Visser, *Deliktereg* (6th edn), LexisNexis, 2010, pp. 36 *et seq.*

202 International Commission of Jurists, p. 19.

203 Supreme Court of Canada, *Mustapha v. Callagan of Canada* [2008] SCC 27 per the Chief Justice, *supra* note 4.

204 Idem *supra* note 16.

205 International Commission of Jurists, pp. 17 and 18.

206 *See* Simons, 2001, p. 907, also for further references.

207 This is by no means only my view; *see also* – *inter alios* – Simons, 2001, pp. 907 *et seq.*

as high as one-third of the global welfare.[208] Such a tremendously adverse impact in itself already calls for bold action. But we cannot close our eyes to the even more important immensity of human suffering if we do not change course right now.

The application of the PETL- and similar formulas in specific cases is not easy. After all, not all of the criteria just mentioned point in the direction of liability. At the end of the day, application of the formulas in a specific case almost unavoidably calls for "value judgments."[209] The US judge Learned Hand rightly observed that "a solution always involves some preference, or a choice between incommensurables."[210] Put differently, the formula provides quite some manoeuvring room for lawyers and particularly for judges. My impression is that quite a few courts only pay lip service to these or similar formulas, but are often just working towards an equitable outcome. Simons put it very eloquently, and in my view convincingly:

> In the end, a determination that an actor is negligent reflects a value judgment at two levels. It expresses the judgment that the actor should have done something different in light of the foreseeable risks of his conduct. It also presupposes value judgments about the relevant advantages and disadvantages of taking such a precaution. The task of conscientiously identifying and clarifying the appropriate value judgments is not easy, but it is unavoidable if negligence is to remain a justifiable ground of tort liability.[211]

Lord Hoffmann has expressed a similar view:

> [. . .] But the balance between risk on the one hand and individual autonomy on the other is not a matter of expert opinion. It is a judgment which the courts must make and which in England reflects the individualist values of the common law.[212]

The same goes for the famous judge Learned Hand in an earlier judgment:

> [. . .] a solution always involves some preference, or choice between incommensurables, and it is thought most likely to accord with commonly accepted standards, real or fancied.[213]

208 *See* Lord Stern on global warming. It's even worse than I thought, *The Independent* of 13 March 2013; Robbins 2012, pp. 10, 11; he also sounds a few optimistic notes, such as the potential of a 50% energy efficiency and links with the ICT revolution that could reduce global emissions by 15% per annum by 2020 (p. 30).

209 *See also* Simons, 2001, p. 916 with a very interesting discussion on the subsequent pages, seen from the angle of various theories (standards).

210 In *Conway v. O'Brien*, quoted in *Prosser, Wade and Schwartz' Torts*, 2010, p. 147.

211 Simons, 2001, p. 935.

212 In *Tomlinson v. Congleton BC* [2003] UKHL 47.

213 *Conway v. O'Brien*, 111 F2d 611, 612 [1940], reversed by the US Supreme Court, 312 U.S. 492 [1941]; *see also* Dobbs, 2000, p. 347.

This "value/result-oriented" approach is not necessarily a blessing. It follows that much will depend on the preferences of the judge who has to decide a specific case and his inclination to take a bold or a timid stance. But it arguably is more of a blessing than a curse. After all, it provides a legitimation for courageous judges to render judgments that are well in line both with the predominant legal doctrine[214] and the urgent demands of our time. The value-oriented approach will not change the position of more traditional judges; but we cannot vest our hope in these.

It follows that

(a) at the end of the day, it largely comes down to value judgments;
(b) it is high noon;
(c) actors should act reasonably;
(d) the dangerousness of the activity and the foreseeability of the looming damage play an important role.

A reasonable person should abstain from activities that, together with those of others, are going to ruin the planet. That view as such will probably not be challenged by many. Possible opponents will argue that things are not so easy, as no single State or enterprise is ruining the world. They would have a point, but I do not think that this argument undermines my 'case'. I wonder whether one could reasonably argue that contributing to global catastrophes is legally acceptable. The following may serve as a (further) underpinning of my point.

The dangerousness of the activity is a somewhat more difficult issue. If we perceive GHG emissions from a *global angle*, they are obviously dangerous. Whether they are dangerous if we judge each *single actor* (the majority of States and all enterprises) is arguably less obvious. One could argue both ways, largely along the lines of the debate in the realm of causation. Realistically speaking, the climate change scene hardly changes if we would ignore the emissions of a single actor (with the exception of, say, the top ten emitters). The global catastrophe will be by and large the same, irrespective of whether these single actors reduce their GHG emissions to zero or do not curb them at all. On the other hand, if none of these actors (or only a small proportion) reduced their emissions, global devastation could not be avoided. That would be manifestly unsatisfactory. As already mentioned, it would imply that the law would not be up to dealing with the real challenges of our time. Although there are arguments for the latter position, I do not think that it is correct.

Our starting point should be the – probably unchallengeable – legal proposition: the greater the risk of harm to others, the greater the burden or duty to avoid injury.[215] That submission does not yet solve the problem. But a brief excursion to traditional tort law cases may serve to overcome the minimal contribution hurdle, as I hope to illustrate.

---

214 *See* for a further elaboration Spier, 2012, particularly Chapters 10 and 12.
215 *Prosser, Wade and Schwartz' Torts*, 2010, p. 148.

According to Van Dam, "fantastic probabilities" may suffice to label an act or omission as wrongful, particularly so in personal injury cases.[216] A similar point is made by Rogers in relation to personal injury cases.[217] The latter is an important finding, given the huge amount of human suffering if we let matters in this field proceed unchanged. A few examples, borrowed from case law, may illustrate that remote chances or an unlikely course of events may suffice for establishing wrongfulness.

First a German case.[218] A child was brought by her mother to the defendant for routine medical investigation. She got a vaccination with her mother's consent. The child subsequently developed a serious disease. The plaintiff based its claim on defective treatment and insufficient explanation of the risks. The parties disputed the frequency of the harm (between 1: 5 million and 1: 750.000); the Court found no more precise clarification necessary, given that the frequency was anyway very low. The Court held that

> The decisive factor for determining whether there is a medical duty to advise is not a particular degree of risk, and especially not a particular statistic; it is whether the risk concerned is specifically attached to the operation and on the realisation of the risk it would be especially burdensome to the patient's lifestyle [. . .]. The Senate therefore adheres to the view that in principle even extremely uncommon risks of this kind must be explained.[219]

A recent English case (*Rolls Royce Industrial Power (India) Ltd v. Cox*)[220] also seems to second the view that a very small contribution may suffice. I have described this case in detail in paragraph 1.3.14.

This judgment seems to suggest that very small contributions may suffice, if I understand correctly seen from a causation and from a wrongfulness (duty of care) angle.[221] Put differently, the '*de minimis* doctrine' is not necessarily a serious obstacle. This submission is reinforced, I think, by the strong language of Lord Justice Kay: "undoubtedly justified the finding that this was not a *de minimis* case."

216 Van Dam, 2006, no. 608-2.

217 *Winfield & Jolowicz on Tort*, 2010, pp. 155 and 156. That matches my own experience over the last more than 15 years in the Supreme Court of the Netherlands.

218 BGH 15.02.2000, NJW 2000, 1784 VI ZR 48/99. Translation apparently by B.S. Markesinis, to be downloaded from the website of the Institute for Transnational Law of the School of Law of the University of Texas.

219 A case about liability for acid rain (BGH 10 December 1987, III ZR 220/86) seems to point in a different direction. According to Koch *et al.*, the acid rain case could be distinguished from climate change "if, as we assume here, it can be shown that the gases are distributed evenly into the atmosphere and therefore that every molecule that is emitted, irrespective where it actually comes from, contributes at least marginally to the greenhouse effect and thereby to the rise in temperature and its consequences" (Koch *et al.*, *in* Lord *et al.*, 2012, p. 401). I wonder whether the latter argument is very convincing, let alone whether it suffices to convince the German Supreme Court to deviate from the acid rain judgment.

220 [2007] EWCA Civ 1189. *See*, more generally, *Clerk & Lindsell on Torts*, 2010, nos. 2-39 *et seq.*

221 *See* the judgment *Rolls Royce Industrial Power (India) Ltd v. Cox.*

The following example may serve as a further underpinning of my submission. In many legal systems it is accepted that when a person knowingly participates in wrongdoing by others, he is liable in full.[222] The following example, borrowed from the Commentary on PETL, may illustrate this point. A group of ten protesters resolves to destroy P's crop. Nine of them go into the field to cut down the plants. One of them (D1) remains outside to watch for the police. While one of the other nine are in the fields, one of them knocks down P and injures him. D1 is – jointly with the others – liable for the damage to the crop. Whether he and the others are also liable for P's personal injury depends on the scope of their agreement with a view to the concerted action.[223]

This case is not about a *chance* but about (the serious threat of) real damage. In these instances, all participants are liable in full, irrespective of whether they have caused any (part of the) loss themselves or not. It is open to debate how far one can stretch this concept. It may, or may not, encompass demonstrations by huge(r) masses, assuming that each participant knows and accepts that (major) harm may be done to other people or property. But let us take the example of a person with a violent record (PV), who has actively participated in a series of demonstrations in the past. There is serious reason to believe that in the days to come a new and very violent and massive demonstration will be held. It is likely that PV is going to participate. Let us assume that a group of potential victims seeks injunctive relief against PV, in that he should stay home. It does not seem far-fetched to assume that there is a fair chance that injunctive relief will be granted, despite the fact that it is unlikely that any specific person of the potential victims will be hit by PV. If injunctive relief would not be granted, and if massive damage would be caused by the group of protesters, there is a fair chance that PV will be held liable for at least part of the loss, assuming that he is one of the very few arrested by the police (and by the same token known to the victims). Victims are probably not under an obligation to prove that PV caused *their* loss.

In the field of tort law, there are many more examples of cases where liability based on fault/negligence was established, despite the fact that the chance of materialization of damage was remote. That goes for Dutch law,[224] but equally for the law of numerous other countries.

A Finnish case may serve as the example par excellence. A child falls from a rock, the tourist attraction of the country. The rock houses a castle, a UNESCO monument. The government did not warn visitors against the inherent dangers of the place. According to the Supreme Court, the government was liable because it did not warn about the danger,

222 *See*, by way of example, Art. 9:101, para. 1 PETL. According to Rogers – one of the leading English experts in the realm of tort law – this view is "relatively uncontroversial" (European Group on Tort Law, 2005, p. 145). *See* for recent developments in US law, *Prosser, Wade and Schwartz' Torts* 2010, pp. 374 *et seq.*

223 European Group on Tort Law, 2005, pp. 146-147.

224 *See* C. van Dam, *Zorgvuldigheidsnorm en aansprakelijkheid*, Deventer, Kluwer, pp. 107 *et seq.*; Hoge Raad 8 January 1982, *NJ* 1982/614; Hoge Raad 6 November 1981, *NJ* 1982/567.

which was well possible "in a simple and inexpensive way."[225] Prima facie not very surprising, but the mere fact that no accident had happened for many consecutive years seems to suggest that the danger was minimal, given the sheer number of tourists.

The facts and the legal setting of the cases discussed before are fundamentally different from the facts relevant to climate change. First, in the German "defective treatment" case, it was undisputed that the *entire* loss was brought about by a specific occurrence. To that extent, the case is incomparable with cases of very minor contributions of the respective actors in the realm of climate change. But there is also a striking resemblance. A very remote chance materialized and gave rise to an obligation to pay damages. In the climate change setting the *chance/probability* is not remote at all, but the *individual contributions* are marginal, or at best small. If materialization of a *very unlikely risk* suffices for *compensation*,[226] it would be unsatisfactory if a very *small contribution* to occurrences that *will affect* the globe at large would not suffice for preventive purposes, particularly for injunctive relief. But even if there would be legal obstacles for *injunctive relief*, I quite strongly believe that marginal contributions are in themselves not a legal obstacle for labelling the GHG emissions of a specific State or enterprise wrongful. Otherwise, there would be no effective remedy, as required by a series of – *inter alia* – human rights conventions.[227] That would be unacceptable, now the stakes in the arena of climate change are of a very different magnitude, compared with the often relatively trivial losses in the cases just discussed. So I quite strongly believe that minimal contribution is not an obstacle for establishing wrongfulness in the setting of climate change.

1.4.5 Wrongfulness: The Obligations of the Respective Players

It follows that minor or even marginal contributions to the global problem should not be an obstacle to quantifying the GHG emissions of a specific State or enterprise as wrongful. That in itself does not mean that they *are* wrongful. To answer the question whether that is the case, the bonus pater familias standard, as elaborated in the PETL, quoted in paragraph 1.4.4, provides the relevant signposts, as discussed in the previous paragraph. Back to the respective criteria elaborated in paragraph 1.4.4. None of them is unambiguous, as the following examples may show. As Lord Rosklill put it:

> There is no simple formula or touchstone to which recourse can be had in order to provide in every case a ready answer to the questions whether, given

225 17 August 2010, KKO 2010, 28, discussed by S. Hakalehto-Wainlo, 'Finnish report', *in* H. Koziol & B. Steininger (Eds.), *European Tort Law 2010*, De Gruyter, 2011, pp. 170 *et seq.*

226 Under Australian law, defendants are required to foresee any risk, however unlikely, provided only that it is not far-fetched or fanciful; Trindade *et al.*, 2007, p. 345.

227 I am not suggesting that the wrongfulness requirement should be abandoned, let alone altogether. All I am suggesting is that the argument may serve as an underpinning of the case of victims for the reasons mentioned below. For the reasons explained in para. 1.1.4, injunctive relief would be the most *effective* remedy.

> certain facts, the law will or will not impose liability for negligence or, in cases where such liability can be shown to exist, determine the extent of that liability. Phrases such as "foreseeability," "proximity," "neighbourhood," "just and reasonable," "fairness" [. . .] will be used from time to time in different cases. But such phrases are not precise definitions. At best they are but labels or phrases descriptive of the very different factual situations which can exist in particular cases and which must be carefully examined in each case before it can be pragmatically determined whether a duty of care exists and, if so, what is the scope and extent of that duty.[228]

First, it is at least open to debate whether there is a relationship (proximity) between, say, a German enterprise and the people in Bangladesh. Arguably not even between "the German people" and a German enterprise based in the very northern part of the country and a victim living in the very south of Germany.[229]
Whether the proximity requirement is met may depend on the size of the GHG emissions. It is quite a step to argue that the requirement is fulfilled in relation to, say, the emissions of a local bakery on the one side and the people of Bangladesh who are forced to leave their drowned homes on the other. One could also cast some doubt in respect of the emissions of, say, a major manufacturer of cars. Personally, I do not think that injunctive relief should founder on this doctrinal reef. If necessary, human rights should be called to aid. I am inclined to believe that violation of a series of human rights could 'replace'" proximity. If that would not be the case, litigators would be well advised to seek for injunctive relief on behalf of the 'right plaintiffs', for example the people at risk in the relevant country or, as the case may be, neighbouring countries.
Except where the costs of precautionary measures – to be dealt with below – are concerned, the signposts point at the need to take swift action. That does not need any elaboration as to the weight of the protected interests affected. Departing from the predominant view, catastrophe will set in if we maintain our wooden-headedness. Massive human and economic suffering will follow, while the environment will be greatly impaired. These interests are obviously protected by law (national and international).
The foreseeability of the damage is so much beyond doubt that this does not need any elaboration either. The mere fact that the predominant view is challenged by sceptics does not carry weight in this respect, as will be explained under the heading of the precautionary principle. One could possibly challenge the view that damage is foreseeable by focusing on the loss brought about by a specific actor. That, by and large, is the same issue already discussed in paragraph 1.4.3.

228 In Caparo, quoted *in Winfield & Jolowicz on Tort*, 2010, p. 165.
229 *See* for a comparable view, though not in relation to climate change, Gloag & Henderson, *The Law of Scotland*, Thomson, W. Green, 2007, no. 27.02. *See also* van Dijk, 2011, pp. 218 and 219.

The dangerousness issue was also discussed in paragraph 1.4.3. For the reasons already mentioned, I do not think that this requirement will turn out to be a serious obstacle.
The only remaining factor is the cost to be incurred in obtaining the reductions required. That factor can best be discussed in the context of the obligations of the respective States and enterprises; see paragraph 1.4.6 below.

1.4.6 The Extent of the Reductions Legally Required

So far, I have concluded that the criteria mentioned in paragraph 1.4.4 are met, although I do admit that this view can be challenged. All in all, I believe that my submission is based on fairly solid grounds.
This brings me to the extent of the reductions *legally* required. That also is one of the vital issues for injunctive relief. If a court would be willing to order such relief, it needs to be very precise and distinct as to the obligations of the defendant.[230,231] The sooner we produce a solid and sufficiently accepted[232] blueprint of the obligations of the respective enterprises and national states, the better.
The answer may depend on the question whether a court would borrow from international law (probably unsettled, so far), human rights law or national tort law. The latter arguably is most advanced in this field.[233] Victims and courts, however, would be best advised to perceive this issue from a broad angle, thus taking all relevant realms of the law into account. This issue was discussed at some length in the previous book.[234] For now, I will only make the following observations, seen from a tort law perspective.

---

230 According to Grossman, courts should not "enjoin all emissions from defendants, since that would destroy those companies," Grossman, 2003, p. 59. I concur, in that it would be a bridge too far to ruin all industry.

231 International courts tend to be cautious and leave quite some manoeuvring room to the States; *see* in more detail Rieter, 2010, pp. 594 *et seq.*

232 Ideally speaking, in international fora, such as the International Commission of Jurists, the International Law Association, leading Environmental fora and the like. Submissions supported by leading experts from various parts of the globe could certainly contribute to an opinio iuris that develops into hard law. Once again, the respective obligations can be based on a series of well-established legal concepts from various parts of the law. One of the leading Indian academics in the climate change arena, Lavanya Rajamani, has advocated stepping back from the blame game, particularly for the past; quite a courageous position from an Indian angle; *see* L. Rajamani, 'India and Climate Change: What India Wants, Needs, and Needs to Do', *India Review,* Vol. 8, No. 3, available at <http://dx.doi.org/10.1080/14736480903116842>, p. 360. I do concur with her to the effect that it may unnecessarily block useful political compromises. But with due respect, I wonder whether this position should also be adopted by judges who have to determine the reductions of the respective countries.

233 *See* for an elaboration Spier, 2012, Chapters 10 and 12.

234 Notably Chapters 10 and 12. *See also* the bold and inspiring submissions by the WBGU report; B. Davey (Ed.), *Sharing for Survival,* Faesta, 2012 and V. Bosetti & J. Frankel, 'Sustainable Cooperation in Global Climate Policy: Specific Formulas for Emission Targets to Build on in Copenhagen and Cancun', A Background Report for the Human Development Report 2011: Sustaining Equitable Progress of the United Nations Development Programme, second draft April 15-25, 2011. *See also* the interesting CRS Report for Congress, R. Meltz, *Climate Change and Existing Law: A Survey of Legal Issues Past, Present, and Future,* 2012, Congressional Research Service, available at <www.crs.gov R42613> (no. R42613); Hunter & Salzman, 2007;

The arguments borrowed from the reasonable person concept and the elaboration in paragraph 1.4.4 quoted above serve two purposes. I have already explained why they could serve as an underpinning for the submission that even marginal GHG emissions may amount to a tort. For similar reasons they provide a legal basis for the submission that States and enterprises should make effective efforts to reduce their GHG emissions. In the *short* term, that rather vague concept may suffice to arrive at major reductions by many States and enterprises.[235] After all:

(a) a not insignificant part of the reductions can be achieved by a more efficient use of energy.[236] If that would be so easy, the very least States and enterprises should undertake is to adopt this type of action to the extent reasonably possible. That would be a win-win situation, as it would save costs and reduce the level of emissions at the same time;
(b) most States and enterprises could change over to carbon-neutral sources of energy, such as solar and/or wind energy.

The downside of the 'effective efforts' approach is that it is rather vague and leaves it to the discretion or willingness of the respective States and enterprises to determine which measures have to be taken. Not every State or enterprise will really make serious efforts to save energy and/or switch to (more) carbon-neutral sources of energy. It would require quite some pressure, either from shareholders, supervisory institutions, NGOs or courts to get the more reluctant players to move. The present state of the world is telling, I think.
If *courts* would be called to determine the meaning of the concept just mentioned, for instance to order injunctive relief, the submission I have just proposed may well turn out less vague than it seems. After all, courts could appoint experts to reduce to concrete

Peel *et al.*, 2012, pp. 245-280, particularly about many local and regional initiatives and Wood, pp. 52 *et seq.* in the context of public trust. *See also* M. Allen *et al.*, 'Scientific Challenges in the Attribution of Harm to Human Influence on Climate', *University of Pennsylvania Law Review*, Vol. 155, 2007, p. 1353; Verheyen, 2005, pp. 350 *et seq.*, particularly pp. 352 and 353; S. Mc-Inerney-Lankford, M. Darrow & L. Rajamani, *Human Rights and Climate Change, A Review of International Legal Dimensions, a World Bank Study,* (2011; A. Agarwal & S. Narain, *Global Warming in an Unequal World, a Case of Environmental Colonialism,* New Delhi, Centre for Science and Environment, pp. 9 *et seq. See* for do's and don'ts in this context S. Manea, 'Defining Emissions Entitlements in the Constitution of the EU Emissions Trading System', *Transnational Environmental Law,* Vol. 1, 2012, pp. 303-323. Sean Stephenson has rightly argued that climate change policies may go at the expense of and by the same token conflict with the right to work, also a human right: S. Stephenson, 'Jobs, Justice, Climate: Conflicting State Obligations in the International Human Rights and Climate Change Regimes', *Ottawa Law Review*, Vol. 52, 2010-2012, p. 155; *see also* S. Jodoin & Y. Saito (Eds.), *Human Rights and the Environment: Key Challenges & Best Practices*, CISDL, 2011. Their point may well be valid and almost certainly the same goes for tensions with other human rights. That does not make things easier, of course. But, given the tremendously adverse impact of climate change, we cannot afford to set modest standards in reducing GHG emissions. First things first! Not surprisingly, Posner and Sunstein contend that 'corrective justice' would mean that innocent people would have to settle the bill. But I do not share their view that 'developed' countries should not be under an obligation to reduce GHG emissions; *see* E.A. Posner & C.R. Sunstein, 'Climate Change Justice', *The Georgetown Law Journal*, Vol. 96, 2008, p. 1565; Kosolapova, 2013, pp. 150, 151, 185, 186.

235 *See* in more detail Spier, 2012, pp. 134 *et seq.*

236 *See* Stern, 2012, p. 30.

figures what could be done. Courts could also urge the defendants to explain what could be done and ask experts to comment on these submissions. But we must anticipate a not overly active cooperation by most defendants, and much will depend on the experts to be appointed. Few will be truly independent, and many will have their hobby horses, as experience in other areas shows. In brief, this approach could contribute to major reductions of GHG emissions, but it is rather unpredictable how it will work.

We should also explore alternative strategies to achieve the bitterly needed global reductions. To that end, the following approach might be more effective. We first have to determine how much GHG emissions must be reduced on a global scale. From there onwards, we must find a fair basis to distribute the burden.

Let us, for the sake of argument, assume that global GHG emissions would have to be reduced by 10% per annum. In that scenario, we would first have to deduct the GHG emissions of the 'developing countries' that are not (yet) under a legal obligation to reduce their emissions and the quantum of the required reductions to be brought about by the emerging economies that is, in my example, below 10%.[237] The only feasible way probably is to work with ballpark figures. These must be provided by leading experts in the field. For the sake of argument, I assume that the ballpark figure of the reductions that developing countries are *not* bound to achieve is 3%. In that scenario, 'developed states' and enterprises would have to reduce their GHG emissions by 13% (10% + 3%) per annum.

One may wonder whether there is a sound underpinning for my submission that 'developed' countries and enterprises in those countries have to assume responsibility for the reductions of 'developing' countries and emerging economies that are not (yet) legally bound to curb their emissions. I admit that an answer in the affirmative could be challenged. But there are quite a few arguments speaking in favour of my submission, I think.

First and foremost, all nations and enterprises know or ought to know the seriousness of the looming threats and the very limited time left to avoid global catastrophe. In such a scenario, responsible actors should assume the 'responsibilities' of other actors that are not *obliged* to change course. To the extent needed, one could argue that the responsibility recognized here rests on an analogous application of the principles referring to

(a) acting in concert;
(b) causing indivisible harm.

The principle supra a (acting in concert) could not be applied without amendment, I think. It is not overly likely that courts are willing to apply this concept lock, stock and barrel. I do not advocate a strict and unlimited application, which would – in many legal

237 By way of example, if countries A-F would only have to reduce their emissions by 5%, the responsibility of the remaining 5% has to be assumed by the 'developed nations'. It is open to debate whether the level of reductions needed can be achieved without the help of 'developing countries.' If not, we have to explore avenues to get them on board. Financial assistance and technology transfer will be needed. See about the latter International Council on Human Rights Policy, Beyond Technology Transfer, Protecting Human Rights in a Climate-Constrained World. The Council rightly observes that these will also be needed for adaptation purposes.

systems[238] – give rise to joint and several liability. For our purpose, a 'mild version' would do, in that it would suffice that a number of actors must assume the 'responsibilities' of other actors with no or lesser obligations.

As a matter of fact, the GHG emissions of all States and enterprises together cause global warming. In quite a few legal systems, those who bring about indivisible harm are jointly and severally liable.[239] Once again, we should not overstretch this concept. More particularly, joint and several liability would be clearly excessive in the context of climate change, also in relation to injunctive relief. After all, it would drive the affected enterprises in the 'developed world' out of business as long as others are not (yet) obliged to curb their emissions. No court will take such a position. For our purpose, a softer variant suffices. Quite a few learned academics will probably object, arguing that in their legal systems actors are not liable for the part of the loss brought about by causes in the victim's sphere, such as natural events (hurricanes, floods and the like). In this reasoning, GHG emissions that do not have to be reduced by the respective States or enterprises belong in the category indicated. Proponents of this defence certainly have a point.[240] But they are not necessarily right, I think. Their defence is based on cases such as losses caused by a series of events, partly human activities and partly events for which nobody is liable. In these instances, the respective human players are unaware of each other's activities. In the setting of climate change, the respective players do know that the aggregate GHG emissions are going to cause serious harm around the globe. They equally know – or ought to know[241] – that a series of countries and enterprises do not (yet) have to reduce their emissions. Both of the scenarios referred to should be distinguished, I think. There are compelling reasons for a restrictive application of the legal doctrine of indivisible harm. Unlike in the field of indivisible harm in general, the respective States and enterprises should not be jointly and severally liable. Instead, they should only assume responsibility for, so to speak, innocent cases (*i.e.* not wrongful emissions by 'developing' countries and enterprises in those countries). That, admittedly, is a kind of legal innovation, but by and large to the benefit of States and enterprises in the rich world, given that they otherwise might be held liable in full.

I do not hide, of course, that my submissions require some courage, as they are by no means accord with what is widely accepted. In a sense, it would be easier to apply well-established legal concepts (acting in concert or indivisible harm), even in the context of climate change. The result would be quite awkward. My approach seems more balanced. In the somewhat longer term, it is in the best interest of all States and enterprises, as it contributes to our goal to prevent global devastation.

My submission has the advantage of simplicity, but it entails some unfairness, not yet discussed. After all, this approach ignores all distinctions. It is indifferent to the fact that

238 *See*, by way of example, Art. 9:101 PETL and *Prosser, Wade and Schwartz' Torts,* 2010, pp. 374 *et seq.*

239 *See*, by way of example, Art. 9:101 PETL and Dobbs, 2000, p. 423. *See* in more detail K. Oliphant (Ed.), *Aggregation and Divisibility of Damage*, Wien; New York, Springer, 2009.

240 *See* Art. 3:106 PETL.

241 Ignorance of the law is rarely accepted as a valid defence.

some States and enterprises have already been active to curb their emissions.[242] It would also serve as a disincentive to get at higher reductions than legally *necessary*.[243] We should try to find a fair solution to this dilemma. The best I can think of is:

(a) to determine a kind of a general standard applying to (developed) countries, along the lines to be elaborated below;

(b) the depart from, say 1980, 1990 or 2000 and adjust the level of emissions that have to be reduced by the reductions already achieved.[244] By way of example, if countries in general would have to reduce their GHG emissions by 10% per annum, a country that has already curbed its emissions as from 1990 by 6% will get a reduction of 6%. The submission *as such* may be fair. the difficulty lies in the starting point: 1980, 1990 or 2000, which may make quite a difference in a concrete case. For now, my imagination falls short to come up with a solid legal underpinning for any of the three years, or any other specific year.

In both scenarios, the easiest would be to attach 'emission rights' to each single country. In principle, the number of inhabitants should be the yardstick. The outcome should, arguably, be adjusted in view of

(a) the countries' capabilities,
(b) historical contributions,
(c) dependence on fossil fuel and
(d) the access to renewable energy.

Thus, some major OECD countries would arguably have to contribute less per capita compared with, say, wealthy countries such as Canada and Italy. Whether or not that would be the case would depend on the weighing of just mentioned criteria. Dubai's historical contributions probably are rather small, but it is not dependent on fossil fuel as it has easy access to renewable energy. This example points to the intrinsic difficulties of any formula that is not very strict. One could easily imagine a strict formula, but these almost inevitably do injustice to quite a few States and enterprises. All in all, we should adhere to a more flexible approach, leaving room for justice to all.

The criteria mentioned above are, as such, far from novelties. But there is not much legal basis for a hierarchy or for giving them a relative weight. The outcome of the discussions in international fora and for example the Annexes to the Kyoto Protocol may also serve

242 Several companies have already reduced their GHG emissions; *see* Stern, Lecture no. 3, p. 16 and Lecture no. 2, 2012, p. 41.

243 Yet States and enterprises would be well advised to curb their GHG emissions as swiftly as possible. Not only for moral reasons, but also to avoid potential liability for damages. After all, it is in the clouds how the law will develop; at some stage, the new insights will be applied retroactively, as is usually accepted.

244 The selection of the year should largely depend on the question of when the serious threats of climate change were sufficiently understood.

as a source of 'inspiration'. But quite frankly, it seems almost impossible to submit a well-underpinned concept, based on wrongfulness or any other private law concept.
Is this to say that the law cannot provide any truly useful basis for determining the obligations of the respective countries? I do not think so. For the time being, the effective efforts approach requires major reductions from at least 'developed' countries. Quite a lot can be easily achieved. After all, all kinds of (more) carbon-neutral equipment are available, and energy efficiency could achieve a not unimportant part of the job.
If courts would come into the picture, they should require a very solid underpinning for the submission that a country is unable to realize more reductions than it is willing to achieve. Thus we can materialize quite significant reductions, but we cannot solve the global problem. The alternative strategy (assumption of the 'responsibilities' of 'developing' nations by the rich countries) aims to address this challenge. In this scenario, the real challenge lies in the obligations of the major group of countries that are entitled to differentiated – *i.e.* lesser – obligations, which is, as we have discussed in paragraph 1.4.3, still rather a terra incognita.
The legal obligations of enterprises arguably are an even more complex issue. First, should we distinguish between enterprises in 'developed' and 'developing' nations, in that they also have differentiated responsibilities? My answer would be an outright no in relation to enterprises with parent companies in a 'developed' country. I realize that this position may serve as an incentive for 'climate change havens' in 'developing' countries, but if that would happen, the answer seems easy: we must pierce the veil and look at reality. More complicated is whether or not 'local' enterprises in 'developing countries' should get a different treatment. In principle, I am inclined to answer this question in the affirmative. Otherwise, the concept of common but differentiated responsibilities would be rather meaningless. This leaves untouched that enterprises manufacturing products for the 'developed' countries[245] and/or for the local wealthy people should arguably have similar obligations as their counterparts in the 'developed' world.
Particularly in the context of enterprises, the final factor of the PETL formula springs to mind: the costs to be incurred to achieve the necessary reduction of GHG emissions. The importance of this factor should not be overstretched. The German Supreme Court is very reluctant to attach any importance to this factor,[246] and rightly so. Van Dam rightly observed that the risk that someone will suffer personal injury generally brings courts to require considerable precautionary measures.[247] Prosser *et al.* observe that the social value of the imperilled interests counts; so does the extent of the harm. Seen from this angle, it is unlikely that the cost issue would outweigh the latter two – all the more so as the costs involved

245 That submission deserves further consideration. After all, those enterprises employ local and often poor people.
246 BGH 29 November 1983, NJW 1984, pp. 801, 802.
247 Van Dam, 2006, no. 807-1.

to come to terms with the looming threats are bearable if we are willing to act in the very short term. So, given the magnitude and seriousness of the threats of climate change that will materialize if we are not willing to reduce GHG emissions to a significant extent and the harm done in that case, a (certain) financial backlash will by no means serve as a valid "excuse" to refrain from taking the necessary steps to curb GHG emissions – all the less so, as the adverse financial impact will be much higher if the looming threats materialize.[248]
The latter submission may not appeal to many economists, politicians, quite a few investors, trade unions and others who confine themselves to short-term views. After all, it points at the urgent need to arrive at major reductions in the very short term. That will indeed go at the expense of the economy, be it, as we have seen, in general to a limited extent. In *concrete cases*, however, the impact may be huge. It is not unrealistic to assume that some – probably just a few – enterprises may go bankrupt if they would have to curb GHG emissions, for example because they already report losses, or at best marginal profits, are engaged in very fierce competition, which means that the margins are very low and similar situations. Under those circumstances, a clear-cut answer may prima facie be obvious. It is arguably impossible to answer this question irrespective of the circumstances.
In the very short term, a balancing of the interests at stake *may* justify the conclusion that these enterprises do not have to reduce their GHG emissions to the extent legally required if they would be unable to afford the measures that would have to be taken. I underscore *may.* It will depend on the circumstances of each particular case whether the "*non habet pecuniam* argument" carries (sufficient) weight. Given that the stakes for society (present and future) are so high, we should be extremely reluctant to lend our ears to pleas of marginal industries to spare them. If at all, the following factors should be taken into account:

(a) the usefulness (for society) of the enterprise and its products;[249]
(b) the question whether or not it would be reasonably possible to cut costs (such as wages and bonuses, travel expenses, all kinds of unnecessary luxury) or whether the enterprise could borrow money;
(c) the expected time span of the losses. If they would last more than at most a few years, the financial argument will not play a role.

My rather rigid position is unquestionably harsh for the enterprises in question and for those dependent on the same (such as employees, creditors, shareholders and the like). But it is justified, I think, bearing in mind that it is an inevitable fact that enterprises that cannot meet their obligations (may) go bankrupt. Seen from that angle, one should not too easily attach decisive importance to the financial argument. More importantly, if we

248 This submission is in line with the view expressed by Trindade *et al.*, 2007, p. 433, though not in the context of climate change. Dan Dobbs convincingly argues that the kind of loss also plays a role; Dobbs, 2000, p. 346.
249 This factor obviously carries some weight. *See*, *inter alia*, Dobbs, 2000, pp. 344 *et seq.*; Van Dam, 2000, p. 174; Hoge Raad 6 april 1990, *NJ* 1990, 573 PAS.

would allow exceptions, too many enterprises will seek for a special treatment, which would mean that our final goal could not be reached. Once again, that will have far more damaging consequences.

This brings me back to the reasonable person-standard, as elaborated above. In the run-of-the-mill cases, remote chances often suffice to establish liability if a loss occurs. That goes particularly for personal injury and damage to the environment. Once again, if we do not change course radically, climate change is going to jeopardize a great many people and the global environment. According to the prevailing view, that is not just a chance but (almost) a certainty. If, ex post facto, a very minor chance may suffice to establish *liability for damages*, a certainty must suffice *ex ante* to create far-reaching legal obligations and to obtain effective preventive remedies – far-reaching insofar as everything possible must be done to avoid unnecessary global damage. With the provisos mentioned above in relation to the common but differentiated responsibilities, it follows that tort law provides a solid basis for the assumption that

(a) we should make the effort necessary;
(b) if still possible, we should avoid that the 2°C threshold will be passed and
(c) if that is no longer possible, we should keep as close to that threshold as reasonably possible;
(d) financial burdens can, if at all, only serve as an excuse to refrain from bold and immediate action in the very short term.

These parameters should be the legal basis for determining the legal obligations of the respective States and enterprises. There are various options to carry them out in more detail and to tailor them to a specific State or enterprise. I have submitted a few suggestions above. The very open and flexible standard of the "reasonable person," as elaborated before, can serve as a sufficiently solid legal basis for a series of far-reaching obligations.

1.4.7 *Other Options*

It is very much open to debate whether the looming crisis can still be warded off by the reduction of GHG emissions only.[250] Other solutions, such as carbon sequestration and geo-engineering spring to mind, but they are rather controversial. Injunctive relief might also come into play in relation to these 'solutions', albeit for a different reason, to avoid that the risks inherent might materialize. If injunctive relief would be sought, therefore, plaintiffs will (probably) not be the likely victims of climate change but potential victims of these measures. To some extent, similar questions to those discussed above will arise.

250 A. Carlin, 'Global Climate Change Control: Is There a Better Strategy Than Reducing Greenhouse Gas Emissions', *University of Pennsylvania Law Review*, Vol. 155, 2007, p. 1401.

But litigation in these fields entails a series of challenges in its own right. It goes beyond the scope of this book to discuss them.
For the time being, I limit myself to the observation that carbon sequestration will give rise to a very delicate balancing. Opinions are divided as to the dangers of sequestration. If I understand correctly, it is still rather unclear whether potential risks will materialize and, if so, if that would or could only happen in exceptional cases or more often. In the latter alternative, it seems quite obvious that injunctive relief would stand a favourable chance. The former alternative (only a risk in exceptional circumstances) would be a nightmare scenario for judges, assuming they would have to balance the interests of global society at large and the interests of a *relatively* small group of people that might be adversely affected by the sequestration, notably often people within the court's jurisdiction. Seen from a moral perspective, the global interests should carry more weight, I think. But it would require quite some courage for a court to accept that and to give lesser weight to the immediate interests of the local people; see also paragraph 1.5.4.

1.4.8 *Positive Developments and Opportunities*

There fortunately are a few positive developments too. A growing number of institutions is becoming increasingly active in relation to climate change. One of them is the World Bank. The Bank is the major development bank, providing huge amounts of money to – mostly – 'developing countries' for specific projects. In that context, it has developed Performance Standards on Environmental and Social Sustainability.[251] These so-called IFC standards are an integral part of IFC's approach to risk management[252] and apply to all projects that have environmental risks and impacts.[253]
Identification of and coping with environmental risks is one of the many useful commitments, required from the Bank's 'clients'. The first standard is about assessment and management of environmental (and social) risks and impacts; the third about resource efficiency and pollution prevention and the sixth about biodiversity conservation. Clients must comply with these standards and also with the applicable national law, including those laws implementing host country obligations under international law. The impact assessment may "recommend alternative (higher or lower) levels of measures, which, if acceptable to IFC, become project- or site-specific requirements.[254] Emphasis is put on avoidance of risks and, if impossible, minimization.[255] This starting point subsequently

251 International Finance Corporation, 2012. The Standards are not primarily about, but they also aim at, climate change; *see* p. 8.
252 International Finance Corporation, 2012, p. 2.
253 *Ibid.*, p. 3.
254 *Ibid.*
255 *Ibid.*, pp. 6, 10 and 22. In its report Development and Climate Change, A Strategic Framework for the World Bank Group, mention is made of the importance the Bank attaches to climate change; *see* pp. 2 *et seq.*

gets softened, I think. The risks and identification process "will consider the emissions of GH gases, the relevant risks associated with a changing climate and the adaptation opportunities."[256] The Performance Standard outlines

> a project-level approach to resource efficiency and pollution prevention and control in line with internationally disseminated technologies and practises. In addition, this Performance Standard [3, JS] promotes the ability of private sector companies to adopt such technologies and practices as far as their use is feasible in the context of a project that relies on commercially available skills and resources.[257]

Further down "financially feasible resource efficiency and pollution prevention principles" reappear.[258] As to GHG, the following goes:

> In addition to the resource efficiency measures described above, the client will consider alternatives and implement technically and financially feasible and cost-effective options to reduce project-related GHG emissions during the design and the operation of the project. [. . .][259]

The World Bank initiatives can only be applauded. They certainly will have a positive impact. Not enough, but the World Bank cannot be blamed for the shortcomings of others, of course. But its initiative – and similar important initiatives, such as the Ruggie Principles and the OECD Guidelines for Multi-National Enterprises[260] – are not very explicit about *what* has to be done. More specifically, they do not mention specific reduction targets of GHG emissions and/or requirements to switch to carbon-neutral sources of energy. This is not meant as criticism. After all, this area is still largely a black spot on the legal map. It also highlights the urgent need to determine the obligations of States and enterprises.

1.5 Precautionary Principle

1.5.1 Introduction

The Australian Judge Stein – rightly – labelled the precautionary principle as "a statement of common sense."[261] In the legal arena it has gained, by and large, the status of a

256 International Finance Corporation, 2012, p. 8.
257 *Ibid.*, p. 22. Further down on the same page mention is made of the objective "to reduce project-related GHG emissions."
258 International Finance Corporation, 2012, p. 23 and footnotes 3 and 4.
259 *Ibid.*, p. 24.
260 *See* in more detail Spier, 2012, pp. 37 *et seq.*
261 Quoted by Preston, 2006, p. 41.

commonly accepted imperative all over the globe.[262] It is laid down in a series of international instruments and has been applied by national[263] and international courts. In the realm of environmental degradation, it is embedded in Principle 15 of the Rio Declaration:

> lack of scientific certainty shall not be used as a reason for postponing cost-effective measures to prevent environmental damage.

An Australian Court put it as follows (references omitted):

> Generally, the precautionary principle in its various formulations has been said to be "preventive," and to involve the minimisation of consequential environmental impact, and the taking of remedial action upon evidence of a significant but not necessarily proveable risk of environmental harm."[264]

The Court subsequently turned to the rationale

> an acknowledgement that science and the scientific method have limitations. Because of the limitations, it is unlikely that the full consequences of the impact of a particular act or activity upon the environment can be known in advance.[265]

In another Australian case, Justice Talbot held that

> the element of caution dictates that the Court, as the consent authority, needs to adopt every avenue open to it in order to *minimise any potential risk of an adverse impact* from the proposal *no matter how remotely connected or unlikely the manifestation* of that risk is[266] (emphasis added).

262 Not surprisingly, the discussion is not confined to the legal arena. The principle also holds the attention of experts in the field of law and economics. *See*, also for a wealth of further references, M.R.A.G. Wibisana, *Law and Economic Analysis of the Precautionary Principle*, Maastricht, Datawyse Maastricht, 2008.

263 *See* for much further comparative information Preston, 2006, *e.g.* pp. 26 *et seq.*; L. Kurukulasuriya, *The Role of the Judiciary in Promoting Environmental Governance and the Rule of Law*, 2003, available at <www.inece.org/5thvol1/lal.pdf>, p. 11 (Argentine, Pakistan and Columbia) and for Kenya Kameri-Mbote & Odote, *in* Lord *et al.*, 2012, p. 312; for India Rajamani & Ghosh, *in* Lord *et al.*, 2012, p. 150 and for Australia Abbs *et al.*, *in* Lord *et al.*, 2012, p. 77; J. Zander, *Different Kinds of Precaution. A Comparative Analysis of the Application of the Precautionary Principle in Five Different Legal Orders*, Maastricht, 2009; EU COM(2000) 1, p. 1. *See* for an elaborate analysis M. Haritz, *An Inconvenient Deliberation: The Precautionary Principle's Contribution to the Uncertainties Surrounding Climate Change Liability*, Oisterwijk, BOX Press Publishers, 2010.

264 Quoted by Preston, 2006, p. 53.

265 *Ibid.*, p. 53.

266 *Port Stephens Pearls Pty Ltd v. Minister for Infrastructure and Planning* [2005] NSWLEC 426 (15 August 2005) at p. 56, quoted by Preston, 2006, p. 60; Zander, 2009; Spier, 2012, pp. 61 *et seq.*

Since the Rio Declaration, the principle has appeared in many national and international legal instruments.[267]

I leave aside whether the same standard should be applied in every specific setting. If, *e.g.*, national law sets specific emission standards for industry, many courts may be reluctant to *fully* assess the premises on which these standards are based if the legal provisions as such would be challenged (if such a challenge is possible at all in the country in point). But courts may – and should – be less reluctant if litigation is about *application* of the relevant law. By way of example, the national law of Ruratania urges its chemical industry to have reduced its GHG emissions by 25% in 2030. This requirement is based on the lowest level of submissions presented by leading climate change experts. Courts should not accept this starting point (25% in 2030), given that it blatantly violates the essence of the precautionary principle.[268] That goes a fortiori if an individual defendant would only rely on the submissions by leading experts to justify its modest reductions.

1.5.2 Application to Climate Change

For the purpose of injunctive relief, three issues deserve specific attention:

- the degree of certainty required,
- the relevant starting point for applying the precautionary principle and
- the importance, if any, of cost-benefit analyses.

The first issue is the easiest. It is beyond doubt *that* climate change poses a serious threat to mankind.[269] This is not to say that the sceptics are necessarily mistaken; time will tell. It is not my aim to discredit them. But as a matter of fact, the prevailing view is supported

267 *See* in more detail UNEP, *Training Manual on International Environmental Law*, pp. 30 *et seq.*; Preston, 2006; N. de Sadeleer (Ed.), *Implementing the Precautionary Principle, Approaches from the Nordic Countries, EU and USA*, London, Earth Scan, 2007 and Communication of the European Commission on the precautionary principle, COM(2000) 1; UNEP, T. Iverson & C. Perrings, 'The Precautionary Principle and Global Environmental Change', 2009, available at <www.unep.org/ecosystemmanagement/Portalis/7/Documents/wp15_precautionary%20principle_UNEP.pdf>, with quite some emphasis on climate change. The discussion does not go beyond mentioning a few relevant criteria; it is hugely vague about the application of the principle.

268 This is, in a sense, rhetoric. After all, the 'blatantly' serves as a ground for my submission. Yet, for the reasons mentioned in the text (the urgency of the matter and the magnitude of the harm if we are willing to run a far from remote risk by sticking to the lowest level of reductions advocated by experts in the field), my reasoning is quite sound, I think. After all, the basis for the low level of reductions is by all reasonable standards mistaken as a starting point for determining the obligations of the respective players, as it is too risky. If we have set our standards too low, there is no way back if catastrophe sets in.

269 UNEP is very right that "Climate change is a global threat that satisfies many of the conditions for which a precautionary approach is indicated." I wonder, however, whether it is entirely justified in mentioning that the "global system is immensely complex and relevant data is missing, making it extremely difficult to assign probabilities across forecasting models" (UNEP report, p. 15). The threats *as such* are well understood. But that leaves untouched that there still are relevant uncertainties, as illustrated below.

by such a number of leading and credible experts that the degree of certainty is sufficient, measured by all common standards.[270] Some legal orders require "substantial evidence," which was interpreted by US Federal courts as "more than a mere scintilla of evidence," but "less than a preponderance,"[271] and "such relevant evidence as a reasonable mind might accept as adequate to support a conclusion."[272] It is beyond reasonable doubt that these thresholds are crossed, given the prevailing view among climate change scientists. Defendants can therefore not reasonably invoke the defence that the threats *as such* are still merely hypothetical and/or that they are not serious enough. Given the predominant view, it is even very much open to debate whether this still is an issue for purposes of the precautionary principle.[273] Almost a rhetorical question, I think.

Seen from a legal – and also from a policy and a moral – angle, the real difficulty lies in the uncertainty about the time we still have to come to grips with the threats of climate change. How much time do we still have to ward off the major threats? How much must we have to reduce our GHG emissions and at what pace? In these respects, opinions still diverge not insignificantly. Part of the story probably is that estimates in the recent past were based on – as it has turned out – overly optimistic scenarios, *i.e.* the hope (and expectation) that GHG emissions would decrease or at least that the increase could be kept within the order of a few percent.[274] Quite the contrary has happened. Another explanation probably is the interpretation of scientific data. This point may be illustrated by the following findings.

According to IPCC – an update is upcoming in the second half of 2013 – GHG emissions have to be reduced by 50%-85% by 2050 if our goal is to keep CO_2 concentration within the range of 350-400 ppm and 30%-60% if we aim for a range of 400-440 ppm.[275] This would imply a cut by 40%[276]-95% for the developed countries by 2050.[277] To limit the temperature increase to 2°C above pre-industrial levels, developing countries would have to reduce their GHG emissions in 2020 by 10%-40%. Developed countries (*i.e.* the so-called Annex I countries to the Kyoto Protocol) would have to curb their GHG emissions by 2020 by 25%-40%.[278]

---

270 *See* in quite some detail Zander, 2009, *inter alia,* pp. 133, 231, 290, 313 and 325.

271 *AT&T Wireless PCS Inc. v. City Council of Virginia Beach,* 155. F3.423, quoted by Zander, 2009, p. 325.

272 *Associated Fisheries of Maine Inc. v. Daley,* 127 F.3d.104, at p. 109, quoted by Zander, 2009, p. 325.

273 According to J.K. Hammitt, the precautionary principle is of little use in relation to climate change; *see* J.K. Hammitt, 'Global Climate Change: Benefit-Cost Analysis vs. the Precautionary Principle', *Human and Ecological Risk Assessment,* Vol. 6, No. 3, 2000, pp. 387-398.

274 IPCC and other experts cannot be blamed, of course, for the deplorable fact that wooden-headedness continues to dominate among those wielding the wheel of power and, admittedly, among most 'ordinary people'. I also must confess to belong to the huge class of sinners, in that my life style is not overly sustainable (either). But we should base our policies and legal answers on the most recent findings, thus keeping pace with the ever evolving state of climate change science.

275 IPPC 2007, 'Summary for policymakers', *in Climate Change 2007:* IPPC, 2007, p. 15.

276 That is probably a mistake. Instead, one should read 80%, I assume. *See also* Chapter 13 of the IPPC, 2007, p. 776.

277 Executive summary Climate Change, 2007: Working Group III: Mitigation of Climate Change.

278 Chapter 13, IPPC, 2007, p. 776.

On 19 November 2012, the World Bank issued an alarming report: *Turn Down the Heat.*[279] It provides a gloomy picture of the world's future if we do not change course. It serves as a strong support for the urgent need for far-reaching reductions of GHG emissions. It serves:

> as a rigorous attempt to outline a range of risks, focusing on developing countries and especially the poor. A 4°C world would be one of unprecedented heat waves, severe droughts, and major floods in many regions, with serious impacts on ecosystems and associated services. But with action, a 4°C world can be avoided and we can likely hold warming below 2°C.
>
> [. . .] **Even with the current mitigation commitments and pledges fully implemented, there is a roughly 20 percent likelihood of exceeding 4°C by 2100.** If they are not met, a warming of 4°C could occur as early as the 2060s.[280]

James Hansen and others have issued a warning that:

> if reductions begin this year[281] the required rate of decline is 6%/year to restore the Earth's energy balance, and thus approximately stabilize climate, by the end of this century. If emissions reductions had begun in 2005, the required rate was 3%/year. If reductions are delayed until 2020, the required reductions are 15%/year. And these scenarios all assume a massive 100 GtC reforestation program, essentially restoring biospheric carbon content to its natural level.[282]

It matters which scenario should be our point of reference. If IPCC is right, we are arguably under a slightly less immense and urgent pressure to embark on far-reaching reductions of GHG emissions right now, compared with the submissions by Hansen *et al.* If the latter are right, the situation is extremely urgent and we do not have any time to lose. If there is more than a fairly remote chance that Hansen *et al.* are right, one should depart from their assumptions about the need, urgency and extent of GHG-reductions required.[283] If sufficiently credible and underpinned reports by other leading or credible experts would point at even more demanding measures that have to be taken to ward off the threat of global

279 Why a 4°C Warmer World Must Be Avoided.

280 Executive summary, p. 1.

281 *I.e.* probably 2012.

282 J. Hansen *et al.*, *Scientific Case for Avoiding Dangerous Climate Change to Protect Young People and Nature*, available at <http://pubs.giss.nasa.gov/authors/pkharecha.html>, p. 2. On p. 10, 3.5% per year is mentioned instead of 3% per year. Departing from 50 GtC reforestation, the level of reduction required would amount to 9% per year (p. 10). The authors point at geo-engineering as a possible solution, albeit not in the short term (p. 10).

283 According to the case law of the ECJ and the Court of First Instance, hypothetical chances do not suffice; *see e.g.* Case (T-392/02), *Solvay Pharmaceuticals v. Council*; Case (390/07), *European Commission v. United Kingdom & Ireland*. *See also* Case (T-13/99) *Pfizer Animal Health* (T-13/99).

devastation, the latter findings should be the legal basis for determining what has to be done and when.[284] This position can be reconciled with the emerging view that we can and should be demanding where vital interests of society at large are at stake.[285] Admittedly, this submission is still a bit vague. So is the state of the law, I think. But I quite strongly believe that we cannot afford to run the risk that present and future generations will (or may) suffer grievous and still avoidable losses.[286] So demanding really means: demanding.[287] That is, we should depart from pessimistic scenarios. They should be the legal basis for determining how much GHG emissions should be curbed, by whom and when.

The issue, arguably, is governed by the precautionary principle – arguably, because the present state of affairs is different compared with the common scenarios/risks governed by the precautionary principle. After all, the latter involve scientific uncertainty about the *potential* risks. In the setting of climate change, the risks *as such* are well understood (floodings, natural catastrophes, droughts, increase of diseases, etc.), but the opinion of eminent experts diverges as to the *urgency and the scope of the measures* to be taken. This doctrinal issue does not need to be exhaustively examined. At the very least, the concept of the precautionary principle paves the way for a 'solution'. It could at least be applied per analogiam.

The EU Commission rightly observed that "requirements linked to the protection of public health should undoubtedly be given greater weight than economic considerations."[288] If we seriously believe that irreversible damage is to be avoided,[289] it follows that we should stay on the safe side. That is even more so, if we accept Justice Talbot's submission that

> the element of caution dictates that the Court, as the consent authority, needs to adopt *every avenue open to it* in order to *minimise any potential risk of an adverse impact* from the proposal no matter *how remotely connected or unlikely the manifestation* of that risk is[290] (emphasis added).

284 *See* for considerably lower levels of risks that may already provide a compelling need for 'action': G.N. Mandel & J.T. Gathii, 'Cost-Benefit Analysis versus the Precautionary Principle: Beyond Cass Sunstein's Laws of Fear', *University of Illinois Law Review*, No. 5, 2006, p. 1051.

285 By way of example, the EU Commission takes the view that economic considerations carry less weight than health concerns; *see* Zander, 2009, p. 111. Climate change is of a very different, *i.e.* much more urgent, nature than any of the issues contemplated so far; *see also* pp. 214 *et seq.* about the Swedish stance. In the US, precaution is much more a requirement (*i.e.* ten times as much) if the interests of infants and children are at stake; Zander, 2009, p. 313. They obviously are in the context of climate change.

286 The EU Communication of 2000 rightly stresses that "the risk measure should not be limited to immediate risks but equally applies to 'situations' in which the adverse effects do not emerge until long after exposure"; Commission of the European Communities, o.c., p. 18.

287 I do not express any personal view on the urgency of the matter. After all, that is far beyond my expertise. More generally speaking, lawyers and courts cannot express any of their own views on this issue.

288 Commission of the European Communities, 2000, o.c., p. 20. Further down (p. 24) it refers to two judgments by the ECJ (C-157/96 and C-180/96, both of 5 May 1998).

289 *See e.g. BGP Properties Pty Ltd v. Lake Macquarie City Council* [2004] 138 LGERA 237 at p. 262, quoted by Preston, 2007, pp. 56 and 57.

290 *Port Stephens Pearls Pty Ltd v. Minister for Infrastructure and Planning* [2005] NSWLEC 426 (15 August 2005) at p. 56, quoted by Preston, 2007, p. 60.

The latter two submissions, which I endorse in the context of climate change, imply, among other things, that we should depart from pessimistic scenarios.
The moral and legal imperative to avoid unnecessary colossal global losses and to embark on significant reductions of GHG emissions is even more obvious if Hansen *et al.* would be right, or if there would be a sufficiently fair chance that they might be right that the price for our unwillingness to effect major reductions by 2005 means that we have to double our efforts right now and that the efforts needed will multiply by more than three if we wait another seven or eight years. If that gloomy assumption is correct, cost–benefit analyses no longer carry any weight, or, perhaps more precisely, only point at the urgent and indisputable need to embark on far-reaching action straightaway.
The counterargument is that we would thus run the risk that we would unnecessarily and prematurely make sacrifices for scenarios that may never materialize. The argument carries some weight, but not enough, I think. First, the prevailing view that climate change poses very serious threats to humanity and that we should try to avoid passing the 2°C threshold unavoidably implies that quite a few measures have to be taken anyway. It may, at some stage, prove true that we have realized more stringent reductions of GHG emissions than strictly necessary in that specific period of time. Even so, our efforts would still have not been pointless or wasted. They will pay back at a later stage, because at that stage society can afford a lower level of reductions. More importantly, let us assume that we would lend our ears to the proponents of this counterargument. I do not deny that they *may* be right, but there is also a real chance that they are mistaken. In that scenario, unnecessary, irreversible and colossal damage will occur. We simply cannot afford to run that risk.
Another and related line of thought points at the economic price that has to be paid if we embark on more reductions than necessary to keep climate change under 'reasonable' control. This argument falls under the umbrella of the prevailing view that some kind of balancing of cost and benefits is required,[291] in that there should be *some* proportionality between the cost of the measures taken and the benefits those measures can bring. For the reasons mentioned above, I do not think that this analysis would lead to a different conclusion.
If we *would have* to forego the greater part of economic activity and/or return to the standard of living of, say, the middle ages, most people would understandably believe that the price would be much too high.[292] But that, fortunately, will not be necessary; far from that. Quite affordable steps are still possible, as follows from the Stern report.[293] That being the case, short-term and relatively modest cost cannot carry more weight than a sufficiently serious chance that well over a hundred million people – and in the longer term well over

291 *See* for a wealth of information and references Zander, 2009, at various places.
292 Admittedly, I might reluctantly concur, although I realize that a billion people or so are still worse off and that this number will drastically increase if we do not achieve far-reaching reductions of GHG emissions.
293 *See* about Stern's more recent views, para. 1.4.4.2.

a billion – will suffer very serious damage.[294] Justice Talbot is right that "extreme caution" is required in case of significant potential harm to the health of human beings and vegetation.[295]

Departing from the precautionary principle, as just advocated, it is certainly conceivable that steps must be taken that will prove unnecessary. It may even happen that some enterprises may be unable to survive if they have to reduce their GHG emissions. This, in turn, will adversely affect people, *e.g.* because they lose their jobs; those concerned may be confronted with lower social security benefits or may otherwise be jeopardized. We should not close our eyes to this adverse impact, of course. But this cannot be nor should be an argument to take optimistic scenarios as one's starting point. If we chose that alternative, we would overlook that much more painful steps will have to be taken if the 'worst-case scenarios' materialize and if we would have based our measures on more optimistic scenarios. In the latter alternative, far-reaching *additional* steps will have to be taken, not only in terms of much higher reductions of GHG gases, but also for effective adaptation measures. Even if one would be willing to ignore the massive and unnecessary human suffering that would occur in this alternative – and I cannot think of any decent reason why we should[296] –, if worst-case scenarios would materialize, the adverse impact on the economy will be colossal, both in absolute terms as in comparison with what a more pro-active stance would have provided.

There have been other instances, of course, where similar questions had to be answered, though in a far less urgent and far less threatening setting. The Supreme Court of India, one of the most progressive and courageous Courts in the world, put it as follows:

> though the leather industry is of vital importance to the country as it generates foreign exchange and provides employment avenues it has no right to destroy the ecology, degrade the environment and pose a health hazard.[297]

Justice Pearlman has observed that "the greenhouse issue should not outweigh all other factors relevant to a determination of whether or not to grant consent."[298] I agree, of course.

294 *See* again Zander, 2009, pp. 111, 132, 148 and 313, although not in relation to climate change.

295 *Nicholls v. Director General of National Parks and Wildlife* [1994] 84 LGERA 397 at p. 419, quoted by Preston, p. 48. Justice Talbot makes his point in relation to a consent about the handling and processing of materials. I cannot think of any valid reason why a similar line of thought would not be appropriate in the context of injunctive relief to stop excessive GHG emissions. Extreme caution means that one has to depart from worst case scenarios.

296 *See* for a similar view, *inter alios,* Mandel & Gathi, 2006, pp. 1046 and 1047.

297 *Vellore Citizens Welfare Forum v. Union of India per Kuldip Singh J.*, AIR [1996] SC 2715, quoted by Preston, p. 62. *See also Conservation Union of South Australia v. Development Assessment Committee and Tuna Boat Owners Association (No. 2)* [1999] SAERDC 86; *see* further Preston, 2007 pp. 54 *et seq.*, pp. 65 and 66; Commission of the European Communities, COM(2000) 1, pp. 20 and 21.

298 *Greenpeace Australia Ltd v. Redbank Power Company Pty and Singleton Council* [1994] 86 LGERA 143, quoted by Preston, 2007, pp. 45 and 46.

Otherwise, industrial activity and even all life would come to a standstill. In the case in point, other beneficial effects – such as, comparatively speaking, lower GHG emissions and the conversion of a waste product into a usable one – carried more weight. McClellan CJ held that consideration of the precautionary principle

> does not preclude a decision to approve an application in any cases where the overall benefits of the project outweigh the likely environmental harm. However, care needs to be taken to determine whether appropriate and adequate measures have been incorporated into such a project to confine any likely harm to the environment.[299]

1.5.3 The Burden of Proof

An important issue is the burden of proof in case injunctive relief is sought to obtain reductions of GHG emissions of a specific defendant. Who has to prove that it is really high noon (*i.e.* we have to act right now), that the required level of reductions should be X% and not, *e.g.*, 0.5 X and, if relevant at all, the feasibility (either technically or economically) of curbing GHG emissions to the extent legally required? There is at least some basis for the submission that the burden of proof lies on the shoulders of the defendant. This point is often made in relation to a similar issue, *i.e.* who has to prove that a new factory or a new process does not impair the environment. The same reasoning could be applied to existing activities.

According to the Judicial Handbook, "those engaging in a potentially damaging activity should have the burden of establishing the absence of environmental harm."[300] The same should obviously go for environmental harm coupled with tremendous human suffering, the toll of climate change.

An Australian Court held that

> [i]t is well accepted in the literature, and it stands to reason, that the proponent needs to satisfy us that the development would be ecologically sustainable. In the matter before us, is the proponent called upon to prove this, only when the applicant has proved, on the balance of probabilities, that there is a threat of serious irreversible damage to the environment? That cannot be the case. [...][301]

299 *BGP Properties Pty Ltd v. Lake Macquarie City Council* [2004], 138 237, quoted by Preston, 2007, pp. 57-59.

300 Shelton and Kiss, 2005, p. 23. A much more lenient approach (the balance of probabilities) was adopted by Stein J in *Leatch v. National Parks and Wildlife Service* [1993] 81 LGERA 270, quoted by Preston, 2007, p. 42.

301 *Conservation Council of South Australia v. Development Assessment Committee and Tuna Boat Owners Association (No. 2)* [1999] SAERDC 86, quoted by Preston, p. 54. The judgment was upheld by the full court of the Supreme Court of South Australia; Preston, 2007, p. 56.

The Supreme Court of India held that

> The "Onus of proof" is on the actor or developer/industrialist to show that his action is environmentally benign.[302]

I do realize, of course, that one could argue that the above does not come into play in relation to the overwhelming majority of potential defendants because their contributions are too small and, by the same token, are not a threat for the world, the environment or humanity. That argument was already discussed at some length in paragraph 1.4.4. It primarily plays a role in the realm of causation and wrongfulness.[303] If those hurdles cannot be taken, there is little – if any – basis for injunctive relief anyway.

1.5.4 Geo-Engineering and Carbon Sequestration

Similar questions arise in the context of carbon storage and/or geo-engineering,[304] albeit (also) for a different reason. I do not express any personal view on these options. But experts are rather critical. See, *e.g.* the British Royal Society.[305] It would be hazardous, if not irresponsible, to rely upon this kind of solutions, particularly geo-engineering. After all, those solutions
(a) May not work and
(b) May cause immense (additional) harm to humanity and/or the environment.

If my understanding is correct, the uncertainties and particularly the risk of carbon storage are of different magnitude, that is to say lower, compared with geo-engineering. The number of people adversely affected, if worst-case scenarios with respect to carbon storage materialize, is probably considerably lower than the number of climate change victims if we fail to come to grips with the looming threats. But that is not necessarily a sufficient reason to opt for this kind of storage. That, in the end, is a matter of balancing and requires in-depth knowledge and understanding of the cost (also and particularly in human and environmental terms) of using and not using this technology; see paragraph 1.4.7.

---

302 *Vellore Citizens Welfare Forum v. Union of India*, AIR 1996 SC 2715; *see also* the same Court, *AP Pollution Control Board v. Prof. M V Nayadu*, AIR 1999 SC 812, quoted by Preston, 2007, pp. 65 and 66 and *Narmada Bachao Andolan v. Union of India*, AIR 2000 SC 3751.

303 As explained above, a *threat* of wrongful acts or omissions will suffice.

304 *See* about the latter IPCC, 2005.

305 The Royal Society, 'Geoengineering the Climate, Science, Governance and Uncertainties', 2009, available at <http://royalsociety.org/uploadedFiles/Royal_Society_Content//policy/publications/2009/8693.pdf>. *See also* ETC Group, Geopiracy, the Case against Geoengineering.

1.5.5 Fracking

In the course of the last five years or so, 'fracking' has become a hot issue. If my understanding is correct, it aims to serve two purposes:

(a) It makes a few countries, such as the United States, less dependent on the import of carbon fuel;

(b) It could reduce the level of GHG emissions.

Opinions are very much divided on the safety of this new technique and equally on its environmental impacts. If used on a (semi-)global scale, fracking *may* seriously jeopardize many people and the environment; the aggregate losses may be colossal.[306] There is an **emerging line of thought that it might have an adverse impact on climate change.**[307]
If fracking would not entail a major global environmental risk, its threats come closer to carbon storage than to geo-engineering. If it involves very serious risks, probably, countries and enterprises are ill-advised to engage in this activity. See in more detail paragraph 1.4.7.

1.5.6 Conclusion

For many, my submission that we have to depart from sufficiently credible worst-case scenarios may not be immediately convincing. I may be helpful to draw parallels with other instances where the precautionary principle was invoked, such as potential health or environmental risks on a considerably lower scale. In many of those instances, "better safe than sorry" was readily accepted.[308] The view that the precautionary principle compels States and enterprises to play safe in cases of relatively minor risks with a very low level of certainty is difficult to reconcile with the idea that it would not compel them to make effective efforts to avoid very substantial likelihood of grievous global catastrophes.
Finally, a report issued by the European Environment Agency in early 2013 shows that there have been many instances of "false negatives" being neglected. Conversely, there are

306 *See* about the risks: The Royal Society and The Royal Academy of Engineering, 2012; M. Schrobe, 'Fracking Outspaces Science on Its Impact', *The Journal of the Yale School of Forestry & Environmental Studies*, 2012; R. Ehrenberg, 'The Facts behind the Frack, Science News', <www.sciencenews.org.>; *The Economist*, 'The Promised Gas Revolution Can Do the Environment More Good Than Harm', <www.economist.com/node/21556249/print>; Executive Summary of Study of the Potential Impacts of Hydraulic Fracturing on Drinking Water Resources: Progress Report, conducted by (US) Environmental Protection Agency, December 2012; Shale gas: energy solution or fracking hell? According to Braden Reddall and Ben Berkowitz, insurers find it difficult to price the risk; they point at one case that was settled in late 2010 for US$ 4.1 million; Analysis: Insurers Find It Tough to Price Fracking Risks, <www.reuters.com/assets/print?aid=USBRE84A13BR20120511>.

307 *See e.g. The Guardian Environment Blog*, 'Fracking: Where's the Debate about Its Climate Change Risks?'.

308 *See e.g.*, S. Lierman, *Voorzorg, preventie en aansprakelijkheid*, Intersentia, 2004 and K.B. Ogilvie, *Applying the Precautionary Principle to Standard Setting for Toxic Substances in Canada*, available at <www.pollutionprobe.org/old_files/Reports/precautionary.pdf> (September 2001).

relatively few incidents of "false positives" where regulatory action has been taken. The report concludes that

> fear of false positives is misplaced and should not be a rationale for avoiding precautionary actions where warranted. False positives are few and far between as compared with false negatives and carefully designed precautionary actions can stimulate innovation, even if the risk turns out not to be real or as serious as initially feared.[309]

The latter findings also support the submission that society at large would be best served if we stay on the very safe side.

1.6 Legal Actions: A Mirage or a Real Option?

In an impressive article, Douglas Kysar, an eminent academic who really cares for the good cause, submits the view that tort law, as it stands, has little to offer to victims:

> Make no mistake: a conceivable set of arguments on behalf of climate change tort plaintiffs *does* exist. The problem, however, is that the winning plaintiffs' direction at nearly every stage of the traditional tort analysis duty would have to encompass "negligence in the air," rather than more particularized relations of responsibility; nuisance would have to be protected as an absolute protection against significant invasions, irrespective of the social welfare balancing; actual cause would have to embrace – at long last – a probabilistic, risk enhancement conception of causation; exceptional measures of apportionment would have to be invoked to address a multiple defendant problem of unprecedented magnitude; proximate cause would have to be interpreted such as the scope of foreseeable harm for greenhouse gas emissions both tracks projections from climate models that stand at the very forefront of scientific inquiry and, in many cases, applies retroactively as a form of imputed knowledge tantamount to strict liability; and harm would have to expanded to include much more by way of anticipatory injury than courts currently recognize.[310]

309 European Environment Agency, 2013, p. 12 (by S. Foss Hansen & J.A. Tickner). They point at "several references and leaked documents" that are said to show that some regulating parties consciously recruited "reputable scientists, media and politicians" to call on if their products are linked to a possible hazard (idem). Marc Le Menestrel and Julian Rode subsequently argue that in "virtually all reviewed cases it was perceived to be profitable for industries to continue using potentially harmful products or operations" (p. 35).

310 Kysar, 2011, p. 44.

Munich Re, one of the leading re-insurers, voices a similar view in a remarkably comprehensive article under the ambitious heading: Climate change and liability – *Everything* you need to know about climate change and liability (emphasis added). Munich Re's message to the world is that it does not cover "direct losses" caused by climate change, if I understand correctly, because it believes that excessive GHG emissions do not give rise to liability; so why should one cover a non-existent risk?[311]

Kysar may, or may not, be right in relation to damages. As to injunctive relief, he points at a series of serious legal challenges. Admittedly, it will not be easy. It will be a bridge too far for quite a few courts. But I do not think that it is impossible, as I have tried to explain in the previous paragraphs.

First, Kysar's submission would mean that the law is irrelevant in relation to the most serious global threat ever, which, if it materializes, will bring about global devastation. That is not a very attractive view (Kysar is the first to admit that). It is not overly likely that judges all over the world will accept his view. Many judges, particularly but by no means only, in so-called 'developing countries', have shown remarkable creativity – and at times courage – to come to grips with major challenges. Secondly, the law has shown enough flexibility to achieve a reasonable outcome in a series of cases. In most instances, those judgments were applauded as they have adapted the law to changing circumstances.

Thirdly, there is no reason why we should look only at tort law, or any other specific realm of the law. On the contrary, we could and should draw from all parts of the law and from well-established legal concepts. Apart from the complex issue of minimal contributions, there is only one other issue that raises a myriad of truly difficult legal questions: the extent of the reductions needed in a particular case. I do not deny that it will be difficult to submit a credible and detailed blueprint on this point. But that is not to say that well-established legal concepts fall short in outlining rather pertinent obligations as – hopefully – was demonstrated before.[312] True, it is work in progress and will probably remain so for some time. We desperately need outstanding academics from various disciplines – not only lawyers – to discuss the respective obligations in depth. Their efforts will be criticized, and they will make enemies; far-reaching obligations are not popular, but is must and it can be done. I do not think this is unattainable, seen from a legal angle.

To many, my submissions may sound as the acme of legal activism. In a sense that is arguably true. But it is submitted that there are stronger and more persuasive arguments for the view that courts should try as hard as they can to provide solutions. According to the prevailing view in the realm of human rights law, there is a duty to provide remedies against violations of these rights. There is a strong current among leading legal experts and also in the international arena to label climate change as a human rights issue, a view

311 I. Ebert & G. Funke, *Climate Change and Liability – Everything You Need to Know about Climate Change and Liability*, p. 5.

312 *See also* Chapters 10 and 12 *in* Spier, 2012.

I whole-heartedly endorse.[313] Departing from this assumption, victims should have effective remedies before national courts or tribunals.[314] There is an emerging view that States are responsible for violations of human rights by enterprises within their territory and that they have to ensure effective remedies.[315] If causation, or any other doctrinal facet of the law, would be an obstacle in all or most cases, there would be no, let alone an effective, remedy for the most serious violations of human rights ever, violations with more grievous harm to more people than ever witnessed in the history of mankind.

In the view of the High Contracting Parties, concluding the Universal Declaration of Human Rights, "recourse" is essential as otherwise "as a last resort," people will – and in their view seemingly may – (have to) opt for rebellion.[316] I most definitely do not advocate rebellion. Instead, friendly and peaceful judicial redress is preferable and also possible.

1.7 UNJUST ENRICHMENT

Weinbaum has argued that unjust enrichment might be a way to avoid the causation obstacle.[317] The suggestion would certainly deserve serious consideration if injunctive relief were not an option. But it would be a second best option if it will not succeed in realizing the bitterly needed reductions of GHG emissions. Nor will it serve as a sufficiently strong incentive to change course.

1.8 INTERLOCUTORY REMEDIES BEFORE INTERNATIONAL HUMAN RIGHTS COURTS

> *[...] even in inter-State proceedings not automatically involving human rights, State sovereignty is not the only relevant principle. Other relevant principles may be effective protection of the environment [...]as well as preventing irreparable harm to persons.*
>
> *Human rights adjudicators have the inherent authority to use provisional measures.*[318]

313 *See* for an elaboration on this point Spier, 2012, pp. 71 *et seq.*

314 Art. 8 of the Universal Declaration of Human Rights; Art. 7 African Charter on Human's and Peoples' Rights; Art. 13 European Convention on Human Rights; and Art. 25 American Convention on Human Rights. *See also* Art. 22 of the International Commission of Jurists (ICJ), Maastricht Guidelines on Violations of Economic, Social and Cultural Rights, 26 January 1997.

315 *See* Ruggie-Principle 1 and for further elaboration Spier, 2012, pp. 40 *et seq.*; *see also* OECD Guidelines for Multi-national Enterprises.

316 *See* the Preamble.

317 A. Weinbaum, 'Unjust Enrichment: An Alternative to Tort Law and Human Rights in the Climate Change Context', *Pacific Rim Law & Policy Journal*, Vol. 20, No. 2, 2011, pp. 429, particularly at pp. 447 *et seq.* As discussed before, Weinbaum is right that causation is a potential stumbling block for the traditional remedies (pp. 441 and 442), but she is mistaken when she casts the difficulty in terms of "an arduous burden of proof." The facts are sufficiently clear and well understood, at least for the purpose of injunctive relief, I think. The legal challenge is the often small extent of the respective contributions.

318 Rieter, 2010, p. 1082.

This chapter is about Human Rights Courts. I leave aside interim relief by other international tribunals, such as the ICJ.[319] For now, it suffices that other international Courts and Tribunals also open avenues for interim relief. Examples are the International Tribunal for the Law of the Sea[320] and the United Nations Human Rights Committee.[321]
For the purposes of this book, interim measures by Regional Human Rights Courts could also play a major role in stemming the tide. As a matter of fact, these institutions do not have an overly impressive track record in relation to the major challenges of our time.[322]
The African Court of Justice and Human Rights[323] has the power

> on its own motion or on the application by the parties, to indicate, if it considers that the circumstances so require any provisional measures which ought to be taken to preserve the respective rights of the parties.[324]

The weakness of the system lies in the 'execution'. True, State Parties must comply with a judgment,[325] but they may choose not to do so. In the latter case the Court shall[326] refer the matter to the Assembly, which "shall decide upon measures to be taken to give effect to the judgment." The Assembly may also impose sanctions.[327] Those sanctions, which *may* be imposed, include denial of the right to speak at meetings, to present candidates for positions within the union, or to benefit from any activity or commitment. Besides, "other measures of a political and economic nature to be determined by the Assembly" may be taken.[328]

319 The ICJ requires an urgent need to prevent irreparable harm; because such harm was not demonstrated, the ICJ did not grant provisional measures in the case of *Argentine v. Uruguay* (Order of 23 January 2007). The ICJ, however, did in the case of *Georgia v. Russian Federation* (order of 15 October 2008). According to Timo Koivurova, the ICJ has only a limited ability to render radical decisions. After all, it has to be careful not to lose the trust of the state community (Koivurova, 2007, p. 296).

320 Arts. 89-95 of the 2009, Rules of the International Tribunal for the Law of the Sea.

321 Rule 86 UN Doc. CCPR/C/3/Rev.3 (1994); *see*, also for other Commissions, HRI, International Human Rights Instruments, HRI/GEN/3 of 6 June 2001.

322 *See* for an in-depth study about provisional measures in International Human Rights settings: Rieter, 2010. She does not specifically discuss climate change, but environmental issues are dealt with at various places. *See* about the case law of the European Court of Human Rights in environmental matters, D.G. San José, *Environmental Protection and the European Convention on Human Rights*, Council of Europe Publishing, 2005. He observes that the Court tries to strike a balance between activism and restraint (pp. 67 *et seq.*); he further points at the pre-eminent emphasis the Court puts on the right to life.

323 A merger of the African Court on Human and People's Rights and the Court of Justice of the African Union; *see* Art. 2 of the Protocol on the Statute of the African Court of Justice and Human Rights.

324 Art. 35, para. 1 of the Statute of the African Court of Justice and Human Rights.

325 Art. 46, para. 1 of the Statute.

326 That obviously is quite an improvement. According to Art. 52 of Rules of the Court of Justice of the African Union, the Court "may" refer to the Assembly which "may" decide on measures. But it seems open to debate whether the present text gives more solace to victims. The mere fact that the Assembly "shall" decide upon "measures" does not seem to entail that they have to be taken, let alone effective measures. This reading is reinforced by the subsequent paragraph about sanctions that "may" be imposed.

327 Art. 46, paras 4 and 5 of the Statute.

328 Art. 23 of the Act of the African Union, read in conjunction with Art. 46, para. 5 Statute.

As to the European Court of Human Rights (ECHR), according to Rule 39 of the Rules of Court,[329] the Chamber, or, where appropriate, its President, may, at the request of a party or any other person concerned, or of its own motion, indicate to the parties any interim measure that it considers should be adopted in the interests of the parties or of the proper conduct of the proceedings before it. Such an interim measure will only be issued if the Court "considers that the applicant faces a real risk of serious, irreversible harm if the measure is not applied."[330] The applicant must state "reasons," *i.e.* "specify in detail the grounds on which his or her fears are based, the nature of the alleged risks and the Convention provisions alleged to have been violated."[331] One should bear in mind that the Court will only deal with "the matter" once all domestic remedies have been exhausted.[332] Once again, the weakness of the system lies in the follow-up. Most countries, belonging to the Council of Europe, tend to obey the Court's judgments, as they should.[333] As a matter of fact, not every country does abide by its judgments. In the latter case the Committee of Ministers will consider "measures to be taken."[334]

Despite the fact that interim measures are binding,[335] they are quite often ignored.[336]

If the Inter-American Court on Human Rights finds a violation of a right or freedom protected by the Convention,[337] it shall – *inter alia* – rule that the injured party "be ensured the enjoyment of his right or freedom that was violated."[338] In "cases of extreme gravity and urgency, and when necessary to avoid irreparable damage to persons, the Court shall adopt such provisional measures as it deems pertinent in matters it has under its consideration."[339]

329 Effective as from 1 September 2012.

330 Practice direction, Requests for Interim Measures, issued by the President of the Court, latest amended on 7 July 2011. The text does not match well with Rule 39 as it seems to overlook that interim measures can also be issued "of its own motion."

331 Practice direction, *ibid.*

332 Art. 35 European Convention on Human Rights.

333 Art. 46, para. 1 European Convention on Human Rights.

334 *See* Art. 46, para. 5, also for further elaboration.

335 Parliamentary Assembly, Preventing harm to refugees and migrants in extradition and expulsion cases: Rule 39 indications by the European Court of Human Rights, by D. Darchiashvili, Doc. 12435 of 9 November 2010 *supra* note 10; *see also* under 44 and 52, both referring to *Mamatkulov and Askarov v. Turkey*; *see* further *supra* note 55, *et seq.* There is no reason to believe that the report does not equally apply to other areas of human rights.

336 *See* the report mentioned in the previous footnote.

337 Strikingly, nine Member States – among them, needless to say, the United States – have not signed the Convention; eleven have not accepted the adjudicatory jurisdiction of the Court (among them the United States and Canada); Inter-American Commission on Human Rights, Reform Process 2012, Reply of the Inter-American Commission on Human Rights to the Permanent Council of the Organization of American States regarding the recommendations contained in the Report of the Special Working Group to Reflect on the Workshop of the IACHR with a View of Strengthening the Inter-American Rights System, dated 23 October 2012, p. 8, *supra* 32.

338 Over the last 30 years, thousands of persons or groups have invoked precautionary measures; *see* Annual Report of the Inter-American Commission on Human Rights 2011, Chapter III C, p. 73, *supra* 10.

339 Art. 63 American Convention on Human Rights.

The State Parties undertake to comply with the Court's judgment. In case of compensatory damages, the judgment "may" be executed in the country concerned in accordance with domestic procedure governing the execution of judgments against the State.[340]

Next to the Court, the parties to the Convention have established the Inter-American Commission on Human Rights (IACHR). Its main function is, *inter alia*, to develop awareness of human rights among the people; to make recommendations for "the adoption of progressive measures in favour of human rights within the framework of their domestic law and constitutional provisions; to take action on petitions."[341] Any person or group of persons or non-governmental organizations may lodge petitions "containing denunciations or complaints of violation of this Convention by a State Party,"[342] albeit it that the remedies under domestic law have to be pursued first and exhausted.[343]

The Convention provides specific provisions for "urgent cases."[344] If no settlement can be reached, the Commission shall draw up a report setting forth the facts and stating its conclusions.[345] If, within three months, the matter has not either been settled or submitted by the Commission or the State concerned to the Court and its jurisdiction accepted, the Commission may "set forth its opinion and conclusions [. . .] for its consideration." Where appropriate, the Commission shall make "pertinent recommendations and shall prescribe a period within which the state is to take the measures that are incumbent upon it to remedy the situation examined."[346]

Precautionary measures are a hot issue for the Inter-American Commission and the Permanent Council of the Organization of American States. The States requested "the definition for the granting of precautionary measures";[347] the same goes for compliance with decisions of the Court and the Commission.[348] As to the former, the IACHR undertook to develop and subsequently publish a digest (*i.e.* a "practice guide") that "systemizes and explains the standards set in precautionary measures and best practises in designing protective measures."[349] The IACHR reveals that "[s]eriousness is analyzed in light of the threat of harm sought to be prevented, and the urgency is determined by a reasonable presumption that the risk or threat involved may materialize, which requires that the response be immediate in order to prevent it";[350] yet it examines each case on

340 Art. 68 of the Convention.
341 Art. 41 of the Convention.
342 Art. 44 of the Convention.
343 Art. 46, para. 1 *supra* a, of the Convention.
344 Art. 48, para. 2 of the Convention.
345 Art. 50, para. 1 of the Convention.
346 Art. 51, paras 1 and 2 of the Convention.
347 *See* Inter-American Commission on Human Rights, Position Document on the Process of Strengthening of the Inter-American System for the Protection of Human Rights, OEA/Ser.L/V/II; Doc. 68 of 8 April 2012, p. 4, *supra* note 13.
348 Idem, p. 7, *supra* 27.
349 Reply, ibid., p. 18, *supra* 63, see also p. 19, *supra* 65 and footnote 63 with further references, nr. 67, 68.; *see also* Directive 2001/80/EC, 2011, p. 19, nrs. 67, 68.
350 Reply, ibid., p. 20, *supra* 81 and 82.

its own merits. The “nature of the risk” plays a role[351]; so does the “continuing nature of the threats” and the follow-up (by the State).[352] In its Annual Report of 2011, the IACHR mentions that it examines three “essential preconditions”: (a) gravity, (b) urgency and (c) the risk of irreparable harm.[353] Strikingly and importantly, the “efficacy of the judicial system” also plays a role.[354] The IACHR further promises to incorporate “the criterion that the beneficiaries can be determined by geographic location or when one can identify the collective group, people, community, or organization to which they belong, and they are in a situation of risk as a result of being part of that community group.” It follows that the measures are not always for the protection of individual persons, but may also protect a collective or a group that is at risk of suffering irreparable harm.[355]

Two cases deserve to be mentioned specifically. First, the case of the Indigenous Communities of the Xingu River Basin in Brazil. The applicants alleged that the life and physical integrity “of the beneficiaries” were at risk due to the construction of a hydroelectric power plant. The IACHR requested the Brazilian government to suspend immediately the licensing process and stop any construction work “from moving forward until certain minimum conditions are met.” Moreover, the

> State must (1) conduct consultation processes, in fulfilment of its international obligations – meaning prior consultations that are free, informed, of good faith, culturally appropriate, and with the aim of these precautionary measures; (2) [. . .]; (3) adopt measures to protect the life and the physical integrity of the members of the indigenous peoples in the voluntary isolation of the Xingu Basin, and to prevent the spread of diseases and epidemics among the indigenous communities being granted the precautionary measures as a consequence of the construction of the [. . .] plant. This includes any diseases derived from the massive influx of people into the region as well as the exacerbation of transmission vectors of water-related diseases such as malaria.[356]

A second case is about the Inhabitants of the community of Omoa, Honduras. The request claimed that the lives and persons of approximately 8,000 inhabitants of Omoa were at risk due to the liquid petroleum storage facility operated by Gas del Caribe within the limits of Omoa city. The location of the facility was allegedly on a geological fault line, vulnerable to natural disasters, thus posing explosion risks. According to a study conducted by the

351 Annual Report 2011, p. 73, *supra* 15.
352 p. 74, *supra* 16.
353 p. 73, *supra* 14.
354 p. 74, *supra* 18.
355 p. 22, *supra* 81–82.
356 p. 77, *supra* 32.

Honduras "prosecution service," the situation could lead to the death of between 103 and 1,400 people. The Commission asked the State

> to take the steps necessary to ensure Gas del Caribe company's effective observance of the environmental regulations and laws in place in Honduras, and to adopt the measures needed to reduce the danger to the lives and persons of the inhabitants of the community of Omoa to an acceptable level.[357]

The second case is less spectacular than the first, not only because the extent of the risk is unclear, but first and foremost because the measure is rather vague. It leaves in the clouds what is meant by "an acceptable level." Be that as it may, the mere fact that the measures were issued is telling and important. It shows that legal action may be very effective in warding off inherent and sufficiently serious risks.[358]
The first case is even more interesting. It can only be applauded that the Court is keen to protect indigenous communities, even if they have chosen "voluntary isolation," despite the fact that, as I assume, the power plant would be of at least some use to society at large. If one is willing to protect a relatively modest community in the given circumstances, it would be difficult to explain why no measures could (and should) be granted in relation to many more potential victims of climate change; victims that often cannot move, unlike probably the Xingu people.[359]
Quite a few Human Rights Courts require, as a general rule, that domestic remedies have been exhausted.

1.9 POLITICAL ARGUMENT

In a better world, truly important issues should be handled by politicians in an adequate way. This goes particularly for the most urgent and most important issue of all: climate change. Most politicians would probably agree. Some of them – mostly, but by no means exclusively, in 'developing' countries and the EU – try hard to realize international agreements. But politicians are, and probably need to be, focused on reaching the best results that can be achieved in the international bargaining process. Even the most ambitious politician, who truly cares for solutions that meet the imperative demands to reduce GHG emissions significantly and very soon, cannot put enough pressure on less ambitious or wooden-headed politicians, or politicians lending their ears to powerful lobby groups of other countries.

357 p. 86, *supra* note 62.
358 Assuming that the State in point is willing to comply with the judgment.
359 To avoid misunderstanding: I am not suggesting that the latter *should* move to a safer place (whether at the expense of the State and/or the plant or otherwise). That is not my point.

Ideally speaking, politicians should *at least* be able and willing to enact useful legislation on a *national* level, *i.e.* legislation that suffices to come to grips with climate change in due time. Some countries, such as Bhutan, Germany, the Maldives, Norway and Switzerland are trying hard to do this. Politicians of quite a few other countries might well be very keen to enact legislation needed to reduce the GHG emissions in their countries to quite some extent. But they face a series of difficulties:

- they are afraid of distorting the level playing field for "their" countries;[360]
- they lack support from the people (voters);[361]
- they face quite some pressure from (parts of) industry or powerful lobby groups to refrain from bold action.

More likely than not, politicians belonging to the class of the "willing" would not mind at all if others would remove the barriers they cannot pull down, although they may not openly admit that. Unfortunately, by no means all politicians are *willing* to stem the tide[362] either because they do not believe in climate change, do not care for the longer term, or due to ineptitude, wooden-headedness or just short-sightedness.

In many countries, the constitution or tradition has it that political choices should be made by the political branches. If the people believe that politicians do a bad job, they can be dismissed (in democratic societies). According to the traditional view, the third (and – ideally speaking – non-political) pillar of the trias politica – the judiciary – should not engage in political issues. Many judges are very reluctant to enter the political arena to deal with – what they perceive as – purely political matters, either because they honestly believe that this is not their task or responsibility or because they do not feel that they are in the best position to assume part of this task even in case of need. In the latter scenario, they assert they do not have the knowledge, the experience or – it is often believed – the backing from the people.

I do not deny that the question *how much* GHG emissions should be curbed entails a not unimportant political component, particularly by the respective players on a national level. The same goes for the interpretation of the common but differentiated responsibilities.

As explained in paragraph 1.4, I have little doubt that the law, as it stands, requires that it must be avoided, to the extent possible, that the fatal tipping point of an increase in

360 That, probably, is also a reason why many enterprises are still reluctant to realize major reductions, despite the laudable good intentions of an increasing number of enterprises. By way of example, Chevron's "7 Principles for Addressing Climate Change" (update April 2012); Chevron observes that "addressing climate change in a meaningful way is a far more complex, long term proposition, requiring implementation of multiple solutions," which – in plain language – seems to suggest that it will not do very much in the short term.

361 It is open to debate whether the argument is very convincing. By way of example, voters in quite a few countries do not support the 'war on terror', but still re-elect the responsible politicians.

362 For them the level playing field, lack of support and the like do not play a major role. They simply believe that they do not have to act and would not appreciate it if they got an 'excuse' to act. The "Tea-Party" proponents may serve as an example.

global temperature with more than 2°C is passed. The difficulty lies in distribution of the burden among the 'developed' and the 'developing' countries. I strongly believe that 'developed' countries have to assume the 'reduction responsibilities' of 'developing' countries and emerging economies that do not (yet) have an obligation to curb their GHG emissions or only to a lesser extent. From there onwards, the picture is less clear. In an ideal world, politicians should attach a very concrete meaning to the common but differentiated responsibilities. But they fall short in meeting their responsibilities.

National politicians *could* distribute the reduction burden of their own country, for instance by requiring their citizens and enterprises to restrict travelling, forbidding use of excessive energy-consuming equipment and the like. They also *could* compel each citizen and enterprise to reduce their GHG emissions with the percentage needed for the country as a whole. If politicians fail to do these things, it is quite a challenge, if it is possible at all, for courts to make these choices. If they have to deliver a judgment, the best they can probably do is to follow the latter approach (every defendant must reduce his emissions with the percentage required for the country as a whole) plus the additional percentage mentioned in paragraph 1.4.6

Admittedly, if injunctive relief is sought, courts find themselves in a delicate position. But that is not to say that courts should abstain. First and foremost, the argument that climate change is a subject inaccessible to courts presupposes that the field is an open space in the law; that the law as it stands – enacted by politicians – is unable to answer, let alone any sufficiently precise, the myriad of fundamental questions that would have to be answered in the context of litigation.

The view that the core question of how much the respective players have to curb their GHG emissions can and should be dealt with only by politicians is mistaken, I think. It is certainly true that the law, as it stands, does not give a crystal clear answer to that question. But that goes for most issues brought before (superior) courts. If the law were always crystal clear, parties would only have to litigate about facts, not about points of law. With some creativity and willingness to think a little bit out of the box, convincing answers can be given to (almost) all legal issues that will arise in the context of climate change litigation. Probably not to decimal places, but that kind of exercise is largely unnecessary, at least in the short term. Ewing and Kysar rightly observe that

> [i]n the common law context [. . .] it is a mistake to assume that judicially manageable standards must already exist in the rule-like form, awaiting application to the case at hand, when one decides whether to invoke the political question doctrine.[363]

363 B. Ewing & D.A. Kysar, 'Prods and Pleas: Limited Government in an Era of Unlimited Harm', *The Yale Journal*, Vol. 121, No. 350, 2011, p. 419.

That view is equally valid for many other legal systems and is not confined to tort law.[364]
As already mentioned before, seen from a legal perspective, the most difficult and delicate question is *how much* GHG emissions have to be reduced by the respective countries and enterprises. We arguably do not have to answer this question in a very pertinent way in the very short term, as long as we are prepared to accept that we must make effective efforts. States and enterprises could and, by the same token, should adopt a very high level of energy efficiency. They ought to abandon unduly energy-consuming activities. More likely than not, these strategies will be insufficient and fall short of arriving at the level of reductions urgently needed. That means that carbon-neutral techniques, such as solar and wind energy, need to be promoted.

Let us assume that all (or most) enterprises in the 'developed' world would be willing to engage in reductions of their GHG emissions at a rate of 20% in the next five years.[365] That would create an enormous demand for solar equipment and the like. It is very much open to debate whether that equipment could be manufactured in such a short time span. There are also other and easier means to achieve reductions, of course. Switching off central heating and lamps when offices or other locations are not in use, reducing the temperature in houses and offices, and a cut in unnecessary travels may serve as examples.[366] But I wonder whether these kinds of measures could bring about the sorely needed reductions. So, for practical purposes, courts (and eventually legislators) *may* still have some time to clarify the outer limits of the respective legal obligations.

Yet it stands beyond reasonable doubt that the technical and other relatively easy measures to curb GHG emissions fall within the outer limits of the enterprises' and States' legal obligations. Given the urgency of the matter and the extremely serious consequences, if we cannot come to grips with climate change in the very near future, it is very likely that a proper interpretation of the law 'as it stands' obliges States and enterprises to go well beyond embarking on 'easy' steps. After all, otherwise human rights, international law and all facets of national laws would only play a role in *relatively* minor and often not overly important cases. I am far from suggesting that evil in individual cases does not matter. It obviously does; happily, it is often redressed by courts. But one cannot deny that global misery – the doom of climate change – is of a very different magnitude. It follows that the latter should

364 Ewing & Kysar seem to adhere to a more pessimistic view, labelling "the Continental tradition" as more bureaucratic and less politically and socially responsive (p. 422). Even if they would be right, this will probably not go for courts in many other parts of the globe.

365 These are ballpark figures. They are probably not exaggerated, bearing in mind that the 'developing' countries have less obligations. If we would depart from the findings of Hansen *et al.*, mentioned before, 20% would fall short in achieving the extent of reductions needed.

366 *See* for many other possibilities ENEP, Bridging Gaps, Watts New? In an English nuisance case (not comparable with climate change, of course) the Court of Appeal held that an injunction to refrain from specific activities is a realistic option, even if that would mean that the defendant would have to close its doors (*Barr & Orrs and Biffa Waste Services Ltd* [2012] EWCA Civ 312, per Coulson, J.).

be given even more attention. At that stage, we reach the hard core of the political argument. For many – politicians, academics and lawyers – it will be quite a step to compel States and enterprises to go (well) beyond technical and other easy means to comply with their obligations. Their concern is understandable, but not justified, I think.

In, I assume, most legal systems, enterprises are held to comply with a variety of rules following from permits and licences. These permits and licences are designed to avoid evil to others. These permits will not be granted if the enterprises' activities pose serious risks to others. If enterprises nevertheless engage in activities (unduly) dangerous to others, courts can step in. If the danger cannot be warded off to the extent needed, then the enterprises will have to refrain from the relevant activity. Courts probably will compel them to do so, if necessary. Admittedly, in these cases there is often a grey zone; the danger can be reduced to quite some extent, but technical or financial means make it difficult to ward it off entirely. In the latter instances, courts may be 'forgiving', close their eyes, hope for the best and refrain from ordering relief.[367] But most courts will not shy away from bold action if the remaining level of danger is too high. One can argue at length about when a danger becomes legally unacceptable. For our purposes, I do not think that this is a very useful discussion. Given that the stakes are tremendously high for society at large, it seems morally and legally imperative to arrive at the reductions *needed* to avoid that we pass the fatal tipping point, even if that would mean that some enterprises cannot survive and that States will have to force the people and enterprises in their jurisdictions to reduce GHG emissions to the extent legally needed; even if that would necessitate painful measures.

This is not to say that judges can or have to prescribe States (and arguably even enterprises) *how* they should achieve the reductions needed. States may choose the means, as long as they achieve the result. If defendants complain that they cannot comply with their legal obligations, courts could remind them that there is a multitude of options, such as switching off heating or lamps when offices are not in use, lowering heating and air conditioning, reduction of travelling, reducing the size of offices, designing buildings and equipment in a more carbon-neutral way and banning all kinds of unnecessary-consuming heating equipment. Defendants may opt for other means to comply with their obligations as long as they arrive at the result. Those who believe that this submission is the acme of (judicial) activism would be well advised to bear in mind that the allegedly overly painful measures we do have to take now will pale compared with the much more demanding measures that will have to be taken if we wait even a few more years.

If one is prepared to accept this reasoning, the 'political argument' should be defused, even – perhaps I should say, also – in the United States. But things are not so easy, I am afraid.

The US Supreme Court has developed a test to determine whether an issue is of a political nature and, by the same token, a forbidden area for judges. For our purpose, the most

367 This submission is based on some knowledge of Dutch practice. I am inclined to believe that it is not only a Dutch disease.

important criteria are a lack of judicially discoverable and manageable standards for resolving the issue; the impossibility of deciding it without an initial policy determination of a kind clearly for non-judicial discretion; an unusual need for unquestioning adherence to a political decision already made.[368] In the climate change context the US Supreme Court has added that the Environmental Protection Agency

> is surely better equipped to do the job than individual district judges issuing ad hoc, case-by-case injunctions. Federal courts lack the scientific, economic, and technological resources an agency can utilize in coping with the issues of this order. [. . .] Judges may not commission scientific studies or convene groups of experts for advice, or issue rules under notice-and-comment procedures inviting input by any interested person, or seek the counsel of regulators in the States where the defendants are located. Rather, judges are confined by a record comprising the evidence the parties present. Moreover, federal district judges, sitting as sole adjudicators, lack authority to render precedential decisions binding other judges, even members of the same court.[369]

For US Federal law this (probably) ends the discussion about injunctive relief against enterprises for "decrees setting carbon dioxide emissions for each defendant at an initial cap," as the Supreme Court summarizes the claim.[370] This leaves untouched that (Supreme) courts in other countries could – and hopefully will – take a more liberal and, with due respect, more courageous and useful stance, whether in situations where national institutions, comparable with the Environment Protection Agency (EPA), are entrusted with setting standards or not. It is shown by the US Court of Appeals, Second Circuit, that such a different approach is certainly possible in the same case.[371] The Court of Appeals observes that earlier case law "set a high bar for nonjusticiability," whereas

> [n]otwithstanding ample litigation, the Supreme Court has only rarely found that a political question bars its adjudication of an issue.

The mere fact that an issue may have political implications does not suffice.[372] Further down, the Court explains that it

368 *Baker v. Carr*, 369 U.S. 186 [1962].

369 *American Electric Power Co, Inc, et al. v. Connecticut et al.*, *supra* 4C.

370 The Supreme Court explicitly left "ajar the door to litigation under state (as opposed to federal) common law for greenhouse gas (GHG) emissions. Some plaintiffs' lawyers are arguing that the decision leaves room for seeking money damages (rather than injunctive relief) even in a federal common law case" (Gerrard, 2011). To say the least, it would be unsatisfactory if damages would and injunctive relief would not be allowable.

371 *Connecticut v. American Electric Power Company Inc*, decided 21 September 2009.

372 *Supra* n. 370, IIA.

> cannot be gainsaid that global warming poses serious economic and ecological problems that have an impact on both domestic politics and international relations. Nevertheless, Defendants' characterization of this lawsuit as implicating "complex, inter-related and far-reaching policy questions about the causes of climate change and the most appropriate response to it" magnifies to the outer limits the discrete domestic nuisance issues actually presented. A result of this magnification is to misstate the issues Plaintiffs seek to litigate. Nowhere in their complaints do Plaintiffs ask the court to fashion a comprehensive and far-reaching solution to global climate change, a task that arguably falls within the purview of the political branches. Instead, they seek to **limit emissions from six domestic coal-fired electricity plants on the ground that such emissions constitute a public nuisance that they allege has caused,** is causing and will continue to cause them injury. A decision by a single federal court concerning a common law of nuisance action, brought by domestic plaintiffs against domestic companies for domestic conduct, does not establish a national or international policy (assuming that emissions caps are even put into place).[373]

The Court subsequently turns to the issue of judicially discoverable and manageable standards. It boils down the defence, observing that federal courts have successfully adjudicated complex common law public nuisance cases for over a century. After a lengthy elaboration of this point, it recalls that federal courts have

> applied well-settled tort rules to a variety of new and complex problems. [. . .] Well-settled principles of tort and public nuisance law provide appropriate guidance to the district court in assessing Plaintiffs' claim and the federal courts are competent to deal with these issues. [. . .] In adjudicating the federal common law of nuisance claim pleaded here, the district court will be called upon to address and resolve the particular nuisance before it, which does not involve assessing and balancing the kind of broad interests that a legislature or a President might consider in formulating a national emission policy. [. . .] Additionally, the fact that this case is governed by recognized judicial standards under the federal common law of nuisance "obviates any need to make initial policy decisions of the kind normally reserved for nonjudicial discretion" and "further undermines the claim that such suits relate to matters that are constitutionally committed to another branch."[374]

373 *Supra* n. 370, IIB1.
374 *Supra* n. 370, IIB2.

I fully endorse the Court of Appeals' view that the law as it stands provides by and large sufficiently discoverable and manageable standards. The Learned Hand formula[375] paves the way, to mention just one doctrine borrowed from US tort law.[376]
Last but not least, the Court of Appeals observes that decisions involving possible limits on carbon emissions have political implications, but

> not every case with political overtones is non-justiciable. It is error to equate a political question with a political case.[377]

As already observed, the US Supreme Court has closed the door for injunctive relief, at least under federal law. Its reasoning is largely based on the idea that the political branch has delegated this issue to the EPA, which means that there is no room left for judicial activity other than a marginal control of the acts of EPA.[378]
For the purposes of this book, it is not helpful to discuss whether or not the view of the US Supreme Court is convincing in the context of the United States (besides, my view on that subject would not have the slightest impact on the Court's case law). My goal is to discuss important issues and to illustrate that a different view is certainly possible. After all, courts in other countries are not bound by US case law. Besides and more importantly, even superior US courts point in different directions. For the reasons mentioned above, I dare to challenge the view that judges are not in a position to adjudicate the question about the extent GHG emissions have to be reduced by (at least) enterprises. But not every court will accept this reasoning. To get the hesitant judges on board, we need to explore other arguments. Let us discuss a few.
Ewing and Kysar have rightly labelled the debate whether the legislative or executive branches have made an "initial policy determination" regarding GHG, and the question what its context might be, as "freewheeling." After all, it presupposes

> that there is just one policy determination to be made regarding greenhouse gases, which alone must dictate whether and how courts can adjudicate public nuisance based on their emissions. [. . .] That another branch is investigating or studying a new response to an emerging problem does not mean that pre-existing common law tort principles – norms always intended to apply to

375 *United States v. Carroll Towing Co*, 159 F.2d. 169 (Second Circuit 1947); *see* for elaboration K.W. Simons, 'The Hand Formula in the Draft *Restatement (Third) of Torts:* Encompassing Fairness as Well as Efficiency Values', *Vanderbilt Law Review*, Vol. 54, No. 3, 2001, p. 901. *See* in more detail para. 1.4.

376 *See* for elaboration Hunter & Salzman, 2007. *See* about a tort law approach also Ewing & Kysar, 2011, pp. 367 *et seq.* and p. 417.

377 *Supra* n. 370, IIB4.

378 *American Electric Power Co, Inc et al. v. Connecticut et al.*, Supreme Court of the United States, decided on 20 June 2011, *supra* 4B. In *Ned Comer et al. v. Murphy Oil USA et al.*, the US Court of Appeals for the Fifth Circuit adheres to the same line.

> evolving circumstances – somehow constitute judicial invasions into political questions.[379]

Further down they express the supposition that

> those who invoke the political question doctrine against common law tort claims do so out of concern less about legally erroneous application of tort doctrine than about legally rightful application.[380]

They hit the mark, I think. Shying away from dealing with claims for injunctive relief also is, in a sense, a judgment in the heart of judicial policy. Proponents of the view that judges who allow claims of plaintiffs in this kind of cases are legal activists overlook that exactly the same point could be made for judges who dismiss claims for injunctive relief in the context of climate change. These critics also fail to notice that there always have been instances where judges had to fill a gap when urgent demands of society so require.

Last, but certainly not the least, unnecessarily high GHG emissions are not only an issue governed by national law. Even if national law would hold that courts must abstain, international law may require them to deal with cases if and to the extent the defendant has violated rules of supranational law. I know that this is a delicate issue; also that not each and every legal system, let alone each and every judge, acknowledges that supra-national law may take precedence. But even in those regrettable instances where national law takes the view that supranational law does not count (to put it bluntly), judges could use the latter as an 'excuse' to deal with cases of an allegedly political nature. A few examples may serve as an illustration of this point.

The ECHR has held on various occasions that States are under an obligation to take adequate action to cope with unduly risky activities that have come to their attention.[381] In the *Öneryildiz* case the Grand Chamber of the ECHR held that

> [t]here is no explicit right in the Convention to a clean and quiet environment, but where an individual is directly and seriously affected by noise or other pollution, an issue may arise under Article 8. Thus, in Powell v. the United Kingdom, (judgment of 21 February 1990, Series A no. 172, § 40), where the applicants complained about disturbance from daytime aircraft noise, the Court held that Article 8 was relevant, since "the quality of [each] applicant's private life and the scope for enjoying amenities of his home [had] been

379 Ewing & Kysar, 2011, p. 381. Further down, they observe that "Indeed, the more one focuses on the instrumental conception of tort law, the more it becomes clear that *all* tort law treads on the territory of the legislative and administrative branches" (p. 386).

380 Ewing & Kysar, 2011, p. 387.

381 *See*, among other cases, *Oneryildiz v. Turkey*; *see* in more detail, Spier, 2012, pp. 71 *et seq.*

> adversely affected by the noise generated by aircraft using Heathrow Airport." Similarly, in Lopez Ostra v. Spain judgment of 9 December 1994, Series A no. 303-C, § 51, the Court held that Article 8 could include a right to protection from severe environmental pollution, since such a problem might "affect individuals' well-being and prevent them from enjoying their homes in such a way as to affect their private and family life adversely, without, however, seriously endangering their health." In Guerra v. Italy (judgment of 19 February 1998, Reports of Judgments and Decisions 1998-I), which, like Lopez Ostra, concerned environmental pollution, the Court observed that "[the]direct effect of the toxic emissions on the applicants' right to respect for their private and family life means that Article 8 is applicable" (§ 57).[382]

The Court recalls that States have a wide margin of appreciation,[383] and subsequently follows its line of thought:

> 98. Article 8 may apply in environmental cases whether the pollution is directly caused by the State or whether the State responsibility arises from failure properly to regulate private industry. Whether the case is analysed in terms of a positive duty on the State to take reasonable and appropriate measures to secure the applicants rights under paragraph 1 of Article 8 or in terms of an interference by a public authority to be justified in accordance with paragraph 2, the applicable principles are broadly similar. In both contexts regard must be had to a fair balance that has to be struck between the competing interests of the individual and of the community as a whole; and in both contexts the State enjoys a certain margin of appreciation in determining the steps to be taken to ensure compliance with the Convention.[384]

For the reasons mentioned above, I quite strongly believe that in the arena of climate change balancing the conflicting interests can only point at far-reaching reductions. Finally, if the political argument were a 'valid' excuse to dismiss injunctive relief in the kind of cases discussed in this chapter, it would mean that there would be no redress for the greatest harm ever done to mankind. That argument may not carry much weight for traditional lawyers who do not look beyond the run-of-the-mill cases. Senior judges ought to have – and often do have – a broader, more encompassing view. They will undoubtedly understand and appreciate that a 'negative position' makes them co-responsible for a global disaster. Once again, Ewing and Kysar put it eloquently: courts grappling with the grievances brought before them

382 Para. 96.
383 Para. 97 with further references.
384 Para. 98.

> give voice to individuals and actors whose grievances have been neglected by the other branches of government, and they structure that voice within the pedigreed, rationalized discourse of law and its principles.[385]

At the end of the day, the future or, if we continue business unchanged, the fate of humanity is in the hands of people: ordinary people, business leaders, politicians and judges. The people at large, captains of industry and senior politicians may care for the good cause, and may be keen to fight the threat of climate change, but as a matter of fact their actions have fallen short, so far. If two of the three pillars of the trias politica shy away from assuming their responsibilities, the other pillar – the judiciary – ought to step in.[386] That is by no means only a normative imperative.

It cannot be denied that we live, in a sense, in what Philippe Sands has eloquently called a lawless world. It is certainly true that not each and every misfortune can be remedied; nor each and every injustice. Appalling poverty, lack of adequate housing, education or medical treatment, massive unemployment that seems to 'write off' whole generations and – in many countries – serious labour accidents or occupational diseases that remain uncompensated may serve as examples. I wonder whether the law is toothless indeed in relation to all or even most of these issues,[387] but that is a huge and rather unexplored topic in its own right. I actually do not know whether most of these abuses could be repaired by legal remedies, given that there is so little case law and doctrine.[388] Too many victims apparently accept their fate.

Be that as it may, we cannot afford such a laissez-faire stance in relation to climate change. The stakes are simply too high. This is not to say that each and every climate change victim is worse off compared with other victims, of course. Such a position would be untenable. But

385 Ewing & Kysar, 2011, p. 421.

386 In quite a few countries, courts at times show remarkable courage in breaking new ground to cope with urgent demands of society or vulnerable people. In the environmental arena, various African courts allegedly show judicial activism, although quite a few other African courts "experience difficulty in resisting the pressure to find in favour of the economic benefits particular projects may bring to the country", M. Faure & W. Du Plessis (Eds.), *The Balancing of Interests in Environmental Law in Africa,* available at <www.pulp.up.ac.za/pdf/2012_04/2012_04.pdf>, pp. 608/611. Further down (p. 612) they point at often inadequate enforcement. See also Kosolapova, 2013, pp. 105 *et seq.*, 117 *et seq.* and 126.

387 *See* about the eradication of poverty *inter alios* T.W. Pogge, 'Eradicating Systematic Poverty: Brief for a Global Resources Dividend', *Journal of Human Development,* Vol. 2, No. 1, 2001; T. Pogge (Ed.), *Freedom from Poverty as a Human Right. Who Owes What to the Very Poor,* Oxford University Press, 2007; J. Spier, *The Law as a Crowbar for the Eradication of Poverty,* Indret 406. The courageous judgement of the South African Constitutional Court in the Grootboom case (CCT 11/00 [2000] ZACC 19) shows that legal remedies are not ruled out, although the follow-up of this judgemnt falls a bit short. India, as so often, does a better job; *see e.g.* Supreme Court Commissioners, Tenth Report of the Commissioners of the Supreme Court, Permanent Shelters for urban homeless populations, the National Report on Homeless for Supreme Court of India, Review of Compliance of State Governments with Supreme Court Orders up to Dec. 31, 2011, WR 196/2001.

388 Losses caused by labour accidents and occupational diseases are covered by (mandatory) insurance in quite a few European countries; in other instances, labour and/or tort law may fill the gap to some extent.

the *cumulative* evil, present and future, that is going to be caused by climate change if all pillars of the trias politica lean back, is so colossal that courts ought, should, no must, take their responsibility. They do not have to rewrite the law; they only have to apply it in an arguably creative way. They have done so in many instances, and these groundbreaking judgments were often applauded for their courage or because they kept pace with the changing demands of society. Not necessarily in the short term, but without the slightest doubt in the somewhat longer term, bold judgments in the realm of climate change will be widely applauded.
If – and I have little doubt that this is the case – the law (treaties, conventions, human rights and a series of national laws) provides a sufficient basis, judges do not have a choice but to firmly take the lead.

1.10 Unorthodox Injunctions: Mirage or Potential Reality?

Injunctive relief against enterprises and national States is by no means a walkover, as demonstrated above. Apart from the legal obstacles that have to be removed, litigation entails quite a lot of practical difficulties. First, potential plaintiffs have to research into which courts might be willing to grant injunctive relief. Secondly, they have to find out the most promising 'targets' for injunctive relief. Given that there are so many enterprises and States that have to change course, it will be difficult to get at injunctive relief against all those who have to reduce their GHG emissions. Thirdly, prospective plaintiffs have to choose the wording of the relief sought very carefully. As things stand now, there are signs that some non-governmental organizations and various others are more motivated by self-promotion than by serving the good cause in giving due consideration to the most promising cases (plaintiffs, defendants, courts and the relief sought). Potential victims would be well advised to bear in mind that bad precedents may not only harm their case, but also, and even more importantly, the cases of many others.
Much would be gained if sufficiently promising parallel tracks could be mapped to arrive at the desirable outcome. Below, I will submit a few suggestions. These submissions certainly need further in-depth analysis. Once again, failures may have an adverse impact on other cases, be it that they will not necessarily hamper injunctive relief against 'ordinary' enterprises and States; after all, the legal basis for the litigation will be very different.
The submissions below are based on the idea that a relatively small number of 'institutions' could bring about a major change. That goes particularly for insurers, banks, major investors and supervisory institutions. After all, insurers could refrain from covering liability (to start with the liability of directors and officers)[389] for climate change or natural

389 Unlike liability for damages of enterprises and national States, the threat of personal liability might serve as a promising vehicle to influence human behaviour. I realize, of course, that damages of the former two are, in my submission, the wrong way ahead, but this submission has not got much support (so far).

catastrophes that are climate change related. If insurers do not refrain from providing such coverage, they may at some stage face claims they cannot pay.[390] Banks may be confronted with debtors unable to repay loans if the debtors are heavily affected by climate change, either due to natural catastrophes or liability. Supervisory institutions, such as central banks, ought to secure that those under their supervision do not engage in unmanageable risks. The latter institutions may be active behind the scenes, but one cannot escape from the impression that many of them let things run their course,[391] possibly because they – too – focus primarily on the short term. Many major investors are pension funds. They should ensure that they are able to pay pensions of those who put their trust in them for the decades to come. By then, their assets will be heavily and adversely affected by the impact climate change is going to have on the economy.

All of them are well placed to take the lead in the climate change debate. Banks, insurers and investors should find ways to lower the risks they are exposed to if they would adhere to business as usual. Insurers and pension funds could use their voting rights as shareholders to bring 'their' enterprises to their senses.[392] Supervisory institutions could urge those under their supervision to refrain from unduly risky engagements. All of them could put pressure on politicians.

Leading insurers seem to have got the point. An increasing number of publications are appearing, and more and more conferences are being held.[393] My impression is that the level playing field and alleged fear for unlawful agreements among insurers[394] about vital policy issues are, thus far, the main reason why they have gone no further than *talking* about the upcoming threats. Supervisory institutions could play a vital role in creating the level playing field.

---

390 Many insurers did not survive the asbestos litigation wave. Major (re)insurers with an impressive track record came close to bankruptcy, due to the enormous number of asbestos claims; *see* for references Spier, 2012, p. 219, footnote 14; Quite a few major (re)insurers would not have survived a second '9/11', if materialized soon after the first. The damage impending because of climate change is of a very different, if considerably more serious, magnitude. The (US) National Association of Insurance Commissioners issued a paper, 'The Potential Impact of Climate Change on Insurance Regulation' (2008); it observes that "[a]lthough market regulation issues and strategies to mitigate the impact of climate change will vary by line of business, solvency-related risks remain central to all insurers and lines of business" (p. 1). A Ceres report *(Climate Risk Disclosure by Insurers, Evaluating Insurer Responses to the NAIC Climate Disclosure Survey* (authored by S. Leurig, 2008) paints a gloomy picture.

391 The (US) National Association of Insurance Commissioners are an example; *see, e.g.*, Climate Change and Risk Disclosure, last updated 29 May 2012; can be downloaded from its website.

392 Major investors are becoming much more active; *see* for a recent and useful initiative Carbon Disclosure Project, The Future of Reporting, CDP FTSE 350 Climate Change Report 2012 on behalf of 655 investors with assets of US$ 78 trillion.

393 *See*, also for further references, the website of the Geneva Association, an impressive insurers' think tank and Spier, 2012, pp. 156 *et seq.*

394 Competition law is often mentioned as an obstacle. Even if it did have that effect, it is difficult to believe that the competition authorities would be unwilling to grant exemptions if they were given to understand that these may be the only way to safeguard the continuity of the relevant companies.

There is a fair chance that these enterprises and institutions will be willing to assume their responsibilities if a credible case could be made that they are under a legal *obligation* to do so. In that – highly preferable – scenario, litigation could be avoided. Given the emerging unease among, at least, insurers and investors, a modicum of legal encouragement may suffice, even more so if those who are at the wheels would understand that they run the risk of personal liability if they seriously neglect their duties. Banks and supervisory institutions may be a harder nut to crack, but the fear of personal liability might also serve as an incentive for their senior officers and directors to act responsibly.

Auditors may also be a promising 'target'. As a matter of fact, they should establish the potential liabilities of enterprises before issuing an opinion about annual accounts. Climate change risks undoubtedly belong to the potential liabilities. Yet it is doubtful whether many auditors are sensitive to these risks. This may give rise to auditor's liability once the unreported liabilities of the enterprises under their 'control' materialize. Unlike what is true as concerns liability for the entire industry or national States, society at large would not be affected if major auditor firms would go bankrupt. So, as such, there is no reason why damages should be ruled out altogether. But it would not do any good either, given that prevention should be the key word.

It is unrealistic to assume that all these players are willing to change course to the extent needed. If they don't, one has to face the question whether injunctive relief might be an option. The first and the most serious stumbling block is probably 'wrongfulness' (negligence). If one would be willing to accept that these players are legally bound to change course, this does not necessarily mean that their obligations can easily be 'enforced' by potential victims. Let us assume that banks are under an obligation to refrain from lending money to excessive GHG emitters, or that insurers should refrain from covering climate change related liability and that they are ignoring this obligation. Is that a tort (or otherwise wrongful[395]) towards other policyholders, to people who have entrusted their money to these banks, or to shareholders, creditors and the like, given that all of them will be adversely affected if these banks or insurers should fail? And, if so, do these banks and insurers owe an obligation towards, say, the people of Bangladesh, because they could play an important role in avoiding the catastrophic consequences of climate change?

Under traditional tort law, the answer is probably in the negative for most of the potential victims enumerated above, but not necessarily for all of them. Similar questions can be posed in relation to supervisory institutions.

People dependent on pension funds probably have a much stronger case. The relationship between them and the pension funds is much closer, if of a contractual nature.

Be this as it may, the law is, as the ECHR has often put it, a living instrument. It is not cast in stone. Even if the law as it stands could not serve as an underpinning for injunctive

395 *I.e.* in the most suitable legal (doctrinal) setting.

relief, courageous judges might be willing to interpret the law in a creative way, which, I think, would be quite *possible*, given that the legal concepts in point are very open and flexible. Once again, if this submission would appeal to others, further research would be necessary.

Let us assume that 'wrongfulness' is not an insurmountable obstacle. In that scenario, injunctive relief seems quite promising. Also in this context, potential plaintiffs should be careful. They should select the most promising courts and the most suitable defendants. And they should pay a lot of attention to the most promising injunctive relief sought.[396]

1.11 Compliance with Licences and the Like

In most instances, enterprises will comply with the conditions set in their national permits or licences. There again, the impact of GHG emissions will increasingly have to be taken into account. In many cases, enterprises will have carried out impact assessments that got the blessing of governmental authorities. Does this bar injunctive relief?[397]

The answer depends on the niceties of the applicable law, of course. It is suggested that, as a general rule, the answer should be in the negative.[398] First and foremost because permits and the like are not an excuse to cause harm to others, let alone grievous harm on a global scale. As a general rule, permits are an obstacle to the public authorities that have granted the same to obligate the enterprise in point to go beyond the terms of the permit. But that should leave civil remedies sought by potentially affected victims untouched.

Not surprisingly, courts in various countries struggle with the permit or licence issue. In a recent judgment per Carnwath LJ,[399] the (English) Court of Appeal discussed the issue at considerable length. The case is about alleged nuisance caused by tipping of industrial and household waste, for which activity a permit was granted. LJ Carnwath first observed that "The common law is at its best when it is simple."[400] He subsequently turns to an analysis of the principles of the law of nuisance. As to the issue at point he arrives at the conclusion that

> [t]here is no principle that the common law should "march with" a statutory scheme covering similar subject-matter. Short of express or implied statutory

396 The submissions in this paragraph go a step beyond the cautious observations in Spier, 2012, pp. 160-161. There are two reasons for this progressively activist position: the ever more alarming forecasts and further development of my own thinking; *see* about the former para. 1.2.

397 Lord *et al.*, 2012, raise the question, but they leave the answer open (p. 35).

398 Art. 7:101 para. 1 *supra* e of the Principles of European Tort Law provides a subtle approach. According to this article liability *can* be excluded. Whether that is the case, depends on a weighing of relevant factors and the conditions for liability. Gerrard, 2013, observes that some decisions have suggested that the fact that a facility has operated under governmental permits might be a defence to a nuisance claim. Under US law, even compliance with a statute is not ordinarily in itself a defence (Dobbs, 2000, p. 572).

399 *Barr & Orrs and Biffa Waste Services Ltd* [2012] EWCA Civ 312.

400 *Barr & Orrs and Biffa Waste Services Ltd* [2012] EWCA Civ 312, *supra* 5.

> authority to commit a nuisance [. . .], there is no basis, in principle or authority, for using such a statutory scheme to cut down private law rights.[401]

In part II of the judgment, Coulson J goes into more detail. He quotes the summary of the Chancellor of the effect of earlier authorities:

> [. . .]it is well established that the grant of planning permission as such does not affect private law rights of third parties.[402]

Even if permits, assessments and the like would serve as an 'excuse' under national law, this would – at least should – not be an obstacle to injunctive relief.[403] First, climate change is not only an issue governed by national law. It is also a human rights issue.[404] An emerging view has it that (multinational) enterprises have to comply with human rights.[405] National permits are no defence under human rights law.

Climate change will greatly and adversely impair the life of many people. By the same token, the right to life comes into the picture. Whatever the legal implications of other human rights may be, the right to life is pre-eminent.[406] It cannot be set aside by other and less compelling considerations. According to the ECHR, the same goes for torture: this is never allowed, not even if vital interests of a State are at stake.[407] If the latter holds true, one can hardly imagine that massive violations of core human rights would be acceptable only because the requirements of a national permit are met. I realize, of course, that each enterprise could argue that it is not 'the cause' of any suffering. For the reasons mentioned in para. 1.3 this argument has some merit, but it should not counterbalance the urgent need to dismiss the 'permit defence'.

1.12 The Vital Role of the Judiciary

The realities of our era are that irresistible pressure may – and probably will – be needed to get things on the move. This points at the vital role judges will have to play to stem the tide. Putting it this way provokes outcries about undesirable judicial activism. Learned senior members of the judiciary have also emphasized the vital role of judges, though in a more subtle way.

401 *Barr & Orrs and Biffa Waste Services Ltd* [2012] EWCA Civ 312, *supra* 77. *See also* Winfield & Jolowicz, 2010, p. 1174.

402 *Barr & Orrs and Biffa Waste Services Ltd* [2012] EWCA Civ 312, *supra*, 46 (ii); *see also supra* note 84.

403 *See Hatton and others v. The United Kingdom*, Judgement of 8 July 2003.

404 *See* in more detail, also for further references, Spier, 2012, pp. 71 *et seq.*

405 *See*, again, the Ruggie Principles and the OECD Guidelines for Multi-National Enterprises.

406 *Streletz, Kessler and Krenz v. Germany.*

407 *Gaefgen v. Germany.*

The UK Supreme Court Justice Carnwath recently observed "general acknowledgement that judges have a vital role to play in the protection of the environment."[408] What follows is telling and less promising. After a few lines about the English system that, if I understand correctly, would work perfectly well in getting the job done, he observes that, in the English context, judges' role is not to create new substantive protections. Carnwath subsequently turns to other countries where – in his view – environmental regulation is "relatively undeveloped." In that setting "courts have had to fill the gap by imaginative interpretation of constitutional guarantees. India is a prime example."[409]

Justice Carnwath is right that the Supreme Court of India – and a few other superior Courts around the globe, which many of us arrogantly believe to be the legal slums of this world – have taken outspoken stances to arrive at solutions to all kinds of abuses. His Lordship is much better acquainted with English law than I am, of course, but I wonder why European and US courts could not provide solutions to the major challenges of our time. It may hold true that this would require some creative interpretation of the law, but I do not think that they would have to overturn the law or well-established legal concepts. It would by no means be a judicial revolution if they would step in.

In reality, my submission is well in line with the way judges have operated in a series of instances. I do not deny that the law is not always crystal clear. If it were, remedies would be *easily* available and litigation conducted around the globe would have been easier. That goes for climate change and also for a series of far less urgent issues. If the legal picture were crystal clear, the great many bookshelves with academic writings about, *e.g.*, climate change would (have) be(en) largely redundant. But as demonstrated in the previous paragraphs, there is a sufficiently solid legal basis for the submission that legal action to fight climate change is a real option.

Back to Carnwath; further down he quotes the former Lord Chief Justice of England and Wales (Lord Woolf) and summarizes the role of the "law" (judges?)

> to ensure informed and transparent decision-making, fair and efficient enforcement of environmental laws, and a fair balance between public and private rights.[410]

The gist of the arguments made in my part of this book is that such a balancing unavoidably ends up in bold judicial action. After all, it is fundamentally unfair to fritter away the future of coming generations and even of the existing generation only to accommodate our short-term hunger for ever higher consumption of quite often unnecessary goods and

---

408 'Institutional Innovation for Environmental Justice', *Pace Environmental Law Review*, Vol. 29, 2012, pp. 556 and 557.

409 *Ibid.*, p. 567.

410 *Ibid.*, p. 558.

services. I realize only too well that the picture is much more complex in poor countries, but that leaves untouched that the 'Western world' should take the lead in reducing GHG emissions significantly. Any other 'balance' would be unfair.
No doubt, many will argue that my daydreams presuppose a sufficiently solid legal basis. After all, even I accept that courts need at least some legal basis for their decisions. I do not deny that one can argue at length about the minimum legal basis needed to arrive at a specific judgment. For now, it may suffice to recall the Principles on the Role of Law and Sustainable Development adopted at a Global Judges Symposium in Johannesburg in 2002. First, the judges "affirm" their commitment to the UN Millennium Development Declaration, which tackles – among many other issues – climate change.[411] More importantly, they express their

> firm conviction that the framework of international and national law that has evolved since the United Nations Conference on Human Environment held in Stockholm in 1972 provides a *sound basis* for addressing the major environmental threats of the day (emphasis added).

The judges subsequently agreed that a series of principles "should guide the judiciary in promoting the goals of sustainable development through the application of the rule of law and the democratic process." One of these principles is "to realize the goals of the Millennium Declaration of the United Nations General Assembly." If taken literally, the latter is a rather meaningless pledge. Conservative judges could – and no doubt will – easily argue that the rule of law is not reconcilable with the concept of interpreting the law in a groundbreaking way. Thus, they would overlook that the latter pledge forms part of a declaration that strongly suggests that
(a) the law as it stands provides a solid basis for legal action and
(b) judges realize that they do not have a choice, but to go out of their way to make the Millennium Declaration a reality.

So we can only pray that Steiner's hope is justified, that all judges, gathered in Johannesburg, really spoke "the same language."[412]
Turning again to the question whether judges can get at the urgently needed judgments on the basis of the law as it stands. I fully second the view of Lal Kurukulasuriya, the chief

411 *See supra* n. 23.
412 Judges Active in Promoting Environmental Law Capacity Building, *Environmental Policy and Law*, Vol. 33, No. 2, 2003, p. 56. Steiner was in those days a judge in the NSW Court of Appeal, the Supreme Court of NSW. He explicitly mentions that four key principles were agreed, among them to realize the goals of the UN Millennium Declaration. *See also* A. Rest, 'Enhanced Implementation of International Environmental Treaties by Judiciary – Access to Justice in International Environmental Law for Individuals and NGO's: Efficacious Enforcement by the Permanent Court of Arbitration', *Macquarie Journal of International and Comparative Environmental Law*, Vol. 1, 2004, pp. 1 *et seq.*

of UNEP Environmental Law Programme. Based on his experience, he observes – in the general context of environmental law – that "more laws are not needed and more institutions are not required. Rather, the application of existing laws within existing institutions by applying a new mindset will elevate environmental considerations into the collective judicial consciousness."[413]

1.13 Final Observations

> *[. . .] The role of the judiciary, individually and collectively, is to be proactive in the delivery of justice.*[414]

In the previous paragraphs, I have discussed a few painstaking issues that will undoubtedly be invoked as defenses in climate change litigation. I do admit that my views are rather outspoken. Part of the academic and judicial community may believe that I have lost sight of reality. They are unfortunately right to the extent that my ideas and those of others promoting the same or similar results are not entirely common ground. The submissions put forward in my contribution to this book breathe an activist tone. My aim, however, is not – and has never been – to set unrealistic goals. But I most strongly believe that, next to initiatives to raise public awareness and to develop technical means to come to grips with the looming threats, legal remedies belong to the very few potentially promising avenues to overcome the deadlocked positions and to achieve results. My submissions are based on a strong moral feeling that we must stem the tide. The urgent need was eloquently phrased by the World Bank:

> The window of opportunity to curb emissions, limit the rise in global temperatures, and build a climate resilience [. . .] is closing.[415]

Legal solutions are close at hand, as I have tried to illustrate.

One could also draw a parallel with judgments about evils against humanity. As explained in the previous paragraphs, injunctive relief is by no means a walkover. Quite a few serious legal obstacles have to be removed. On the short(er) term, litigation may founder on one or more of these obstacles. Once catastrophe has set in, or the threats have become so alarming that they cannot be ignored any longer, courts will understand that they cannot

413 Kurukulasuriya, 2003.

414 The (then) Lord Chief Justice of England, Lord Woolf, quoted by Kurukulasuriya, 2003, p. 7. According to Kurukulasuriya, environmental jurisprudence, especially over the past three or four decades, provides illuminating evidence of the rise of proactive judicial engagement in resolving disputes that require delicate balancing of competing and compelling interests.

415 World Bank Group, Development and Climate Change, p. 21.

abstain any longer. Law will progress, as it has done on many occasions in many fields. When that stage is reached, nobody will understand why injunctive relief and other useful remedies were not an option in our era.

It is in the laps of the Gods how the law will react to those who knowingly choose to fritter away vital interests of society. The reaction may be lenient, but that probably was also the expectation of people like the former Eastern Germany party leader Aegon Krenz and many others. It is certainly not beyond imagination that, at some stage, those who have contributed to the mess and those who provided obstacles to prevent catastrophes will feel the weight of the law.[416] I do not dare to predict whether that will happen and, if so, how and when.

Criminal relief after the event may appeal to the sense of justice of future victims of climate change, but it would be of no practical avail to them.[417] It would be very helpful to all of us and even more so to future generations if devastation could be avoided. We should spare no means to get there. It will require some courage and a little bit of foresight. It may require mapping new avenues, built on well-established legal concepts, national and international legislation and case law. Future generations will be thankful that we have spared them.

Legal strategies will not work, unless judges are willing to deliver courageous judgments. In part, special courts, entrusted to deal with environmental matters, may counter the challenges.[418] The latter issue is discussed in quite some detail in Ulrich Magnus' part of this book. But international environmental courts could also be counterproductive. One cannot avoid the impression that some of these courts try to strike a balance between the obvious challenges of our time and the everyday reality that they will lose support if they, by traditionally conservative standards, engage in judicial activism. The same goes, to a lesser extent, for national courts. Lesser, because they only have to 'deal' with national sentiments. This tension is noticeable from a speech by Lord Justice Carnwath. He first mentions the earlier-mentioned Johannesburg principles:

> We [the judges present] affirm that [. . .] judicial [process] is vital for the implementation, *development* and enforcement of environmental law (emphasis added).

He subsequently underscores the 'development' issue:

> Underlying to this project to my mind are four fundamental beliefs: first, the task of safeguarding the environment for present and future generations,

416 *See*, by way of example, *Krenz, Streletz and Kessler v. Germany*. The tribunals after World War II, the Yugoslavia-Tribunal and the ICC may also serve as a warning.

417 In this book Gerhard Kemp paints a credible picture of the role criminal law could play. It can only be hoped that his submissions will bring those affected to their senses *before* massive global evil is done.

418 *See*, also for further references, P. Bakshi & M. Yadav, 'New Judicial Roles and Green Courts in India', available at <http://inece.org/conference/9/proceedings/08_Scanson.pdf>.

> while providing an acceptable standard of life, is probably the greatest challenge we face in the 21st century; secondly that time is running out; thirdly, that principles which should guide our response to that challenge (sustainable development, precautionary principle, public trusteeship) form a shared pool of knowledge and experience, which is now recognised in one form or another by most of the legal systems of the world; and fourthly, that judges at all levels, national and international, are uniquely placed, as decision makers and opinion formers, to put those principles into practice.

After having raised the rhetorical question of how judges can be impartial "about the future of our environment – an issue in which all human beings have an interest," he concludes slightly less promisingly:

> We do not show partiality by treating them [i.e. the principles of sustainable development] as underlying principles for all our decision-making in the environmental field. But they must be applied within the constraints of our own legal systems. There is always a tension between judicial creativity and judicial conservatism. Our first duty is to the law, as it is. But as early as the 14th century Chief Justice Shareshull insisted: "No precedent is of such force as that which is right."[419]

Lord Carnwath is right. In the realm of sustainable development, there may well be a tension between the law as it stands and the law "that is right." Even I admit that there are limits to judicial activism in this (and other) fields. But given that the stakes are so high (higher than ever before), judges should be willing to break new ground to the extent reasonably possible to ward off catastrophes that would otherwise affect the future and well-being of humanity. Just a daydream? Time will tell.

I do not dare to predict the position that the ECHR will take if it has to deal with a complaint in the realm of climate change. But victims could remind the Court – and other Courts – of its own and wise case law. In a groundbreaking judgment the Grand Chamber discussed how to determine the meaning of the terms and phrases used in the Convention. After mentioning a series of other factors, the Court held:

> Recourse may also be had to supplementary means of interpretation, either to confirm a meaning determined in accordance with the above steps, or to establish the meaning where it would otherwise be ambiguous, obscure or manifestly absurd or unreasonable [. . .]. Since the Convention is first and

419 *The Ethics of judging*, <https://data.iucn.org/dbtw-wpd/html/EPLP-060/section13.html>.

> foremost a system for the protection of human rights, the Court must interpret and apply it in a manner which renders its rights practical and effective, not theoretical and illusory.[420]

The case law of the ECHR illustrates that these are not meaningless words. They could – and should – get a new dimension in the arena of climate change and sustainable development more generally. The winds of change are not in favour of huge emitters.

At the end of a symposium in Johannesburg, Senior Judges adopted a statement; they "affirm" their commitment to the pledge of the world leaders "to spare no effort to free all of humanity, and above all our children and grandchildren, from the threat of living on a planet irredeemably spoilt by human activities (. . .)." They "emphasize that the fragile state of the global environment requires the Judiciary as the guardian of the Rule of Law, to boldly and fearlessly implement and enforce applicable international and national laws, which in the field of environment and sustainable development will assist in (. . .) sustaining an enduring civilization, and ensuring that the present generation will enjoy and improve the quality of life of all peoples, while also ensuring that the inherent rights and interests of succeeding generations are not compromised."[421]

Laudable words. All these and other judges need to do is to translate them into action.

420 *Demir and Baykara v. Turkey*, para. 65, quoted by B.E.P. Myjer, *Mensenrechten zijn niet soft*, Wolf Legal Publishers, 2012, pp. 8 and 9. A similar point was made in the context of climate change by Coen Drion (before he was appointed in the Supreme Court of the Netherlands), *NJB* 2007, p. 2857. This is not to say that Dutch courts would be the most promising courts to test climate change litigation; for obvious reasons I do not want to comment on that point.

421 UNEP News Release 2002/58.

Table of Cases

ICJ

ECHR

ECJ

ECFI

Australia

Canada

Waddah Mustapha v. Culligan of Canada Ltd. [2008] SCC 27 — 24

England & Wales

Barr & Ors v. Bifa Waste Services Limited [2012] EWCA Civ 312 — 86, 96, 97
Chaplin v. Hicks [1911-1913] All ER Rep 224 — 39
Donogue v. Stephenson [1932] AC 562 — 47
Johnston v. NEI International Combustion Limited, Rothwell v. Chemical and Insulating Company Limited and others, Topping v. Benchtown Limited and Grieves v. F T Everard & Sons and others [2007] UKHL 39 — 4
Morris and Hartlepool Co Ltd [1956] AC 552 — 48
Rolls Royce Industrial Power (India) Ltd v. Cox [2007] EWCA Civ 1189 — 25, 51, 61
Sykes v. Minister of Defence QBD 19-Mar-1984 — 38
Woollerton and Wilson Ltd v. WH Smith (Eziot) Ltd [1987] 1 W.L.R. 853 — 35

Germany

BGH 29 November 1983, NJW 1984, 801 — 60
BGH 10 December 1987, III ZR 220/86, NJW 1988, 478 — 51
BGH 15 February 2000, NJW 2000, 1784 VI ZR — 51

India

AP Pollution Control Board v. Prof. M V Nayadu, AIR [1999] SC 812 — 73
Vellore Citizens Welfare Forum v. Union of India, AIR [1996] SC 2715 — 73

Netherlands

Hoge Raad 6 November 1981, *NJ* 1982/567 — 52
Hoge Raad 6 April 1990, *NJ* 1990/573 — 61
Hoge Raad 24 October 1997, *NJ* 1998/257 — 38
Hoge Raad 16 February 2007, *NJ* 2007/256 — 38

South Africa

Grootboom, CCT 11/00 [2000] ZACC 19 — 92

USA

American Electric Power Inc et al. v. Connecticut et al., 564 U.S. ... (2011) — 87
Associated Fisheries of Maine Inc v. Daley, 127 F.3d 104 — 67
AT&T Wireless PCS et al., v. City Council of the City of Virginia Beach, 155 F.3d 423 — 67
Raymond E. Britt v. The Cyril Bath Company et al., 417 F.2d 433 — 13
Ned Comer et al. v. Murphy Oil USA et al. — 16
Connecticut et al., v. American Electric Power Inc et al. — 15, 87, 89

Bibliography

Abbs, R., Cashman, P. & Stephens, T., 'Australian Report', *in* R. Lord *et al.* (Ed.), *Climate Change Liability*, Cambridge, Cambridge University Press, 2012.

Agarwal, A. & Narain, S., *Global Warming in an Unequal World, a Case of Environmental Colonialism* (2nd reprint), New Delhi, Centre for Science and Environment, 2003.

Akkermans, A.J., *Proportionele aansprakelijkheid bij onzeker causaal verband*, Deventer, Tjeenk Willink, 1997.

Allen, M., *et al.*, 'Scientific Challenges in the Attribution of Harm to Human Influence on Climate', 155(6) *University of Pennsylvania Law Review*, 2007.

Allsopp, M., Santillo, D. & Johnston, P., *Plastic Debris in the World's Oceans*. Available at <www.unep.org/.../docs/plastic_ocean_report.pdf>.

Alston, P., 'A Human Rights Perspective on the Millennium Development Goals', Paper prepared as a conference contribution to the work of the Millennium Project Task Force on Poverty and Economic Development. Available at <www2.ohchr.org/english/issues/millennium-development/docs/alston.doc>.

Anderson, K. & Bows, A. Referred to in *James Hansen and the Climate-Change Exit Strategy: Monthly Review*. Available at <http://monthlyreview.org/2013/02/01/james-hansen-and-the-climate-change-exit-strategy>.

Anton, D.K. & Shelton, D.L., *Problems in Environmental Protection and Human Rights: A Human Right to the Environment*. Available at <http://ssrn.com/abstract=187937>.

Bakshi, P. & Yadav, M., *New Judicial Roles and Green Courts in India*. Available at <http://inece.org/conference/9/proceedings/08_Scanson.pdf>.

Barrett, S., *All-Night Negotiations in Doha Produce no Cliffhanger*. Available at <http://yaleglobal.yale.edu/print/8277>.

Blodgett, J. & Parker, L., *Greenhouse Gas Emission Drivers: Population, Economic Development and Growth, and Energy Use*, 2010, Congressional Research Service. Available at <www.crs.gov RL33970>. CSR Report 2.

Blomquist, R.F., 'Comparative Climate Change Torts', 46 *Valparaiso University Law Review*, 2012, p. 1053. Available at <http://scholar.valpo.edu/vulr/vol46/iss4/3>.

Boom W.H. van, 'Comparative Notes in Injunction and Wrongful Risk-Taking', 17(1) *Maastricht Journal of European and Comparative Law*, 2010.

Bosetti, V. & Frankel, F., *Sustainable Cooperation in Global Climate Policy: Specific Formulas for Emission Targets to Build in Copenhagen and Cancun, A Background Report for the Human Development Report 2011: Sustaining Equitable Progress of the United Nations Development Programme.* Second Draft. 15-25 April 2011, <www.hks.harvard.edu/fs/jfrankel/sustainablcopperatnclimate2012.doc>.

Braams, W., Rijn, A. van & Scheltema, M., *Klimaat en recht*, Deventer, Kluwer, 2010.

Brown, L.R., World on the Edge. *How to Prevent Environmental and Economic Collapse*, New York & London, W.W. Norton & Company. Available at <www.earth-policy.org>.

Bulterman, M.K. & Kuijer, M. (Eds.), *Compliance with Judgments of International Courts*, The Hague, Martinus Nijhoff Publishers, 1996.

Busnelli, F.D. & Comandé, G., *in* Spier, J. (Ed.). *Causation*.

Capice, D., 'The Looming Food Crisis', *YaleGlobal Online*, 2011. Available at <http://yaleglobal.yale.edu/print/6855>.

Carlin, A., 'Global Climate Change Control: Is There a Better Strategy Than Reducing Greenhouse Gas Emissions?', 155 *University of Pennsylvania Law Review*, 2007, p. 1401.

Lord Justice Carnwath, *The Ethics of Judging*, 2006. Available at <http://data.iucn.org/dbtw-wpd/html/EPLP-060/section13.html>.

Carnwath, R., 'Institutional Innovation for Environmental Justice', 29 *Pace Environmental Law Review*, 2012, pp. 555, 556-557.

Clapham, A., *Human Rights Obligations of Non-State Actors*, New York, Oxford University Press, 2006.

Clerk & Lindsell (Eds.), *on Torts* (20th edn), 2010, Sweet & Maxwell.

Commission of the European Communities, *Communication from the Commission on the precautionary principle,* COM(2000) 1 final, 2000.

Davey, B. (Ed.), *Sharing for Survival*, FAESTA, 2012.

Deakin, S., Johnston, A. & Markesinis, B., *Markesinis and Deakin's Tort Law* (6th edn), Oxford, Clarendon Press, 2008.

Deng Haifeng, 'Chinese Report', in Lord, R. et al. (Eds.), *Climate Change Liability*, Cambridge, Cambridge University Press, 2011.

Derraik, J.B.G., 'The Pollution of the Marine Environment by Plastic Debris: A Review', 44 *Marine Pollution Bulletin,* 2002, pp. 842-852.

Dölle, M., Mahony, D. & Smith, A., 'Canadian Report', in Lord, R. et al. (Eds.), *Climate Change Liability*, Cambridge, Cambridge University Press, 2011.

Dobbs, D.B., *Remedies*, West Publishing, 1993.

Dobbs, D.B., *The Law of Torts,* West Group, 2000.

Dubats, E., 'An Inconvenient Lie, Big Tobacco Was Put on Trial for Denying the Effects of Smoking: Is Climate Change Denial Off-Limits?', 7(2) *Northwestern Journal of Law & Social Policy, 2012.*

Ebert, I. & Funke, G., *Climate Change and Liability – Everything You Need to Know about Climate Change Liability.*

Ehrenberg, R., *The Facts behind the Frack*, Science News. Available at <www.sciencenews.org>.

Engel, K., 'Damages, Injunctions, and Climate Justice: A Reply to Jonathan Zasloff', 58 *UCLA Law Review Discourse*, 2011, p. 189.

Enneking, L.F.H., *Foreign Direct Liability and Beyond. Exploring the Role of Tort Law in Promoting International Corporate Responsibility and Accountability*, The Hague, Eleven International Publishing, 2012.

European Environment Agency, 'Late Lessons from Early Warnings: Science, Precaution, Innovation', *EEA Report No. 1*, 2013.

European Group on Tort Law, *Principles of European Tort Law, Text and Commentary*, Vienna; New York, Springer, 2005.

Ewing, B. & Kysar, D.A., 'Prods and Pleas: Limited Government in an Era of Unlimited Harm', 121 *The Yale Law Journal*, 2011, p. 350.

Farber, D.A., 'The Case for Climate Compensation: Justice for Climate Change Victims in a Complex World', 2 *Utah Law Review*, 2008, p. 377.

Farouk, D. & Youssef, L., 'Egyptian Report', *in* Lord, R. *et al.* (Eds.), *Climate Change Liability*, Cambridge, Cambridge University Press, 2011.

Faure, M.G. & Nollkaemper, A., 'International Liability as an Instrument to Prevent and Compensate for Climate Change, 26A *Stanford Environmental Law Journal*, 2007.

Faure, M. & Du Plessis, W. (Eds.), *The Balancing of Interests in Environmental Law in Africa*, 2012. Available at <www.pulp.up.ac.za/pdf/2012_04/2012_04.pdf>.

Fridman, G.L.H., *The Law of Torts in Canada* (2nd edn), Carswell, 2002.

Gagné, M. & Moisan, J-M., *Environmental Class Actions and Claims for Health-Related Injuries: Causation and Its Pitfalls*, 2012. Available at <www.mccarthy.ca/article_detail.aspx?id=5977>.

Galand-Carval, S., 'Causation under French Law', *in* Spier, J. (Ed.), *Unification of Tort Law: Causation*, The Hague, Kluwer Law International, 2000.

Geopiracy, *The Case against Geoengineering*. Available at <www.etcgroup.org>.

Gerrard, M.B., *Global Climate Change and U.S. Law*, American Bar Association, 2007.

Gerrard, M.B., 'What Litigation of a Climate Nuisance Might Look Like', *Yale Law Journal, YLJO Essay*, 2011.

Gerrard, M.B., 'Federal Executive Actions to Combat Climate Change', 49 *New York Law Journal*, 2013, p. 249.

Gerrard, B. & Fischer Kuh, K. (Eds.), *The Law of Adaptation to Climate Change. U.S. and International Aspects*, American Bar Association, 2012.

Giesen, I., 'Proportionaliteit in de klimaatdiscussie', *Nederlands Tijdschrift voor Burgerlijk Recht*, 2012/51.

Glazewski, J. & Collier, D., 'South Africa', *in* Lord, R. *et al.* (Eds.), *Climate Change Liability. Transnational Law and Practice*, Cambridge, Cambridge University Press, 2012, pp. 319-348.

Gloag & Henderson, *The Laws of Scotland*, Thomson/W. Green, 2007.

Global Reporting Initiative, *A Resource Guide to Corporate Human Rights Reporting*. Available at <www.globalreporting.org/humanrights>.

Gonzáles Marquez, J.J., 'Mexican Report', *in* Lord, R. *et al.*, *Climate Change Liability*, Cambridge, Cambridge University Press, 2011.

Griss, I., Kathrein, G. & Koziol, H., *Entwurf eines neuen österreichischen Schadenersatzrechts*, Springer, 2006.

Grossman, D.A., 'Warming Up to a Not-So-Radical Idea: Tort-Based Climate Change Litigation', 28(1) *Columbia Journal of Environmental Law*, 2003, pp. 1-61.

Hakalehto-Wainlo, 'Finnish Report', *in* Koziol, H. & Steininger, B. (Eds.), *European Tort Law 2010*, De Gruyter, 2011.

Hammitt, J.K., *Global Climate Change: Benefit-Cost Analysis vs. the Precautionary Principle*. Available at <http://dx.doi.org/10.1080/10807030091124536>.

Hansen, J. *et al.*, *Scientific Case for Avoiding Dangerous Climate Change to Protect Young People and Nature*. Available at <http://pubs.giss.nasa.gov/authors/pkharecha.html>.

Hansen, J., *A New Age of Risk, Mobilizing Science for Sustainable Development: The Sustainable Development Solutions Network*, Presentation at Columbia University 12 September 2012. Available at <www.columbia.edu~jeh1/presentations.shtml>.

Haritz, M., *An Inconvenient Deliberation: The Precautionary Principle's Contribution to the Uncertainties Surrounding Climate Change Liability*, Oisterwijk, BOX Press Publishers, 2010.

Hart, H.L.A. & Honoré, T., *Causation in the Law*, 2nd edn, Clarendon Press, 1990.

Hodge, W.C. *et al.*, *Torts in New Zealand, Cases and Materials* (2nd edn), Oxford University Press, pp. 212.

Hunter, D. & Salzman, J., 'Negligence in the Air: The Duty of Care in Climate Change Litigation', 155 *University of Pennsylvania Law Review*, 2007, p. 1741.

Idso, C.D., *The State of the Earth's Terrestrial Biosphere*, Center for the Study of Carbon Dioxide and Global Change, 2012. Available at <www.co2science.org>.

International Commission of Jurists, *Corporate Complicity & Legal Accountability* (Vol. 3), *Civil Remedies*, Report of the International Commission of Jurists Expert Legal Panel on Corporate Complicity in International Crimes.

International Council on Human Rights Policy, *Beyond Technology Transfer, Protecting Human Rights in a Climate-Constrained World*, Geneva, 2011.

International Finance Corporation, The World Bank Group, *IFC Performance Standards on Environmental and Social Sustainability*, 2012. Available at <www.worldbank.org/climatechange>.

IPCC, *Climate Change 2007. Working Group III: Mitigation of Climate Change.* Available at <www.ipcc.ch/publications_and_data/.../publications_ipcc_fourth_assessment_report_synthesis_report.htm>.

Iverson, T. & Perrings, C., *The Precautionary Principle and Global Environmental Change*, UNEP, 2009. Available at <www.unep.org/ecosystemmanagement/Portalis/7/Documents/wp15_precautionary%20principle_UNEP.pdf>.

Jodoin, S. & Saito, Y. (Eds.), *Human Rights and the Environment: Key Challenges & Best Practices*, CISDL, 2011.

Johnson, S., 'Climate Change and Global Justice: Crafting Fair Solutions for Nations and Peoples', 33 *Harvard Environmental Law Review*, 2009, p. 297.

Jones, M.A. & Dugdale, A.M. (Eds.), *Clerk & Lindsell on Torts* (20th edn), London, Sweet & Maxwell, 2010.

Kadner-Graziano, T., *in* Winiger *et al.* (Eds.). *Digest of European Tort Law.*

Kameri-Mbote, P. & Odote, C., 'Courts as Champions of Sustainable Development: Lessons from East Africa', 31 *Sustainable Development Law & Policy*, 2009.

Kameri-Mbote, P. & Odote, C., 'Kenya', *in* Lord, R. *et al.* (Eds.), *Climate Change Liability. Transnational Law and Practice*, Cambridge, Cambridge University Press, 2012, pp. 296-318.

Kaminskaite-Salters, G., *Climate Change Litigation under English Law*, Maastricht, Universitaire Pers, 2010.

Kamminga, M.T. & Zia-Zarifi, S. (Eds.), *Liability of Multinational Corporations under International Law*, The Hague, Kluwer Law International, 2000.

Klaus, V., *Blue Planet in Green Shackels. What Is Endangered: Climate or Freedom*, Competitive Enterprise Institute 2008.

Knox, J.H., 'Linking Human Rights and Climate Change at the United Nations', 33 *Harvard Environmental Law Review*, 2009, p. 477.

Koch, H-J., Lührs, M. & Verheyen, R., 'German Report', *in* Lord, R. *et al.* (Eds.), *Climate Change Liability*, 2012.

Koivurova, T., 'International Legal Avenues to Address the Plight of Victims of Climate Change: Problems and Prospects', 22 *Journal of Environmental Law and Litigation*, 2007, p. 267.

Kosolapova, E., *Interstate Liability for Climate-Related Damage*, Den Haag, Eleven International Publishing, 2013.

Koziol, H., *in* Spier J. (Ed.), *Causation*.

Koziol, H., *Basic Questions of Tort Law from a Germanic Perspective*, Jan Sramek Verlag, 2012.

Krauss, L.M., 'Deafness at Doomsday', *The New York Times*, 2013. Available at <www.nytimes.com//deafness-at-dommsday.html>.

Kurukulasuriya, L., *The Role of the Judiciary in Promoting Environmental Governance and the Rule of Law*, 2003. Available at <www.inece.org/5thvol1/lal.pdf>.

Kysar, D.A., 'What Climate Change Can Do about Tort Law', 41(1) *Environmental Law*, 2011. Available at <http://digitalcommons.law.yale.edu/fss_papers/3849>.

Leiserowitz, A., Smith, N. & Marlon, J.R., *Americans' Knowledge of Climate Change*, New Haven, Yale University, Yale Project on Climate Change Communication, 2010.

Leurig, S., *Climate Risk Disclosure by Insurers: Evaluating Insurer Responses to the NAIC Climate Disclosure Survey*, A Ceres Report, 2011. Available at <www.ceres.org>.

Lierman, S., *Voorzorg, preventie en aansprakelijkheid*, Antwerpen, Intersentia, 2004.

Linden, A.M. & Feldhusen, B., *Canadian Tort Law* (6th edn), Lexis Nexus Butterworth.

Lord, R., et al., *Climate Change Liability. Transnational Law and Practice,* Cambridge University Press, 2012.

Loubser, M. & Midgley, R. (Eds.), *The Law of Delict in South Africa,* Oxford University Press, 2009.

Luntz, H. *et al., Torts, Cases and Commentary,* LexisNexis; Butterworth Australia, 2009.

Macrory, R. (Ed.), *Principles of European Environmental Law,* Groningen, Europa Law Publishing, 2004.

Maguire, R. & Lewis, B., 'The Influence of Justice Theories on International Climate Policies and Measures', 8(1) *Macquarie Journal of International and Comparative Environmental Law*, 2012.

Mandel, G.N. & Gathi, J.T., 'Cost-Benefit Analysis versus the Precautionary Principle: Beyond Cass Sunstein's Laws of Fear', 5 *University of Illinois Law Review*, 2006, p. 1037.

Manea, S., 'Defining Emissions Entitlements in the Constitution of the EU Emissions Trading System', 1(2) *Transnational Environmental Law*, 2012, pp. 303-323.

Mank, B.C., 'Civil Remedies', in Gerrard, M.B. (Ed.), *Global Climate Change and U.S. Law*, 2007.

Mc-Inerney-Lankford, S., Darrow, M. & Rajamani, L., *Human Rights and Climate Change: A Review of the International Legal Dimensions.* A World Bank Study. Available at <www.worldbank.org>.

Meltz, R., *Common-Law Climate Change Litigation after American Electric Power v. Connecticut*. Congressional Research Service, 2011. Available at <www.crs.gov R14496>.

Meltz, R., *Climate Change and Existing Law: A Survey of Legal Issues Past, Present and Future*, Congressional Research Service 2012. Available at <www.crs.gov R42613>.

Mickelson, K., 'Beyond Politics of the Possible? South-North Relations and Climate Justice', 10 *Melbourne Journal of International Law*, 2009, p. 411.

Miller, M.E., 'The Right Issue, the Wrong Branch: Arguments against Adjudicating Climate Change Nuisance Claims', 109 *Michigan Law Review*, 2010, p. 257.

Moréteau, O. & Pellerin-Rugliano, C., *in* Winiger *et al.* (Eds.), *Digest of European Tort Law.*

Myjer, B.E.P., *Mensenrechten zijn niet soft*, Oisterwijk, Wolf Legal Publishers, 2012.

NAIC, *The Potential Impact of Climate Change on Insurance Regulation*, 2008. Available at <www.naic.org>.

Neethling, J., Potgieter, J.M. & Visser, P.J., *Deliktereg* (6th edn), LexisNexis, 2010.

Nicholas de Sadeleer (Ed.), *Implementing the Precautionary Principle: Approaches from the Nordic Countries, EU and USA*, London, Earth Scan, 2007.

Ogilvie, K.B., *Applying the Precautionary Principle to Standard Setting for Toxic Substances in Canada*, 2001. Available at <www.pollutionprobe.org/old_files/Reports/precautionary.pdf>.

Oliphant, K., *in* Winiger *et al.* (Eds.), *Digest of European Tort Law.*

Oliphant, K. (Ed.), *Aggregation and Divisibility of Damage,* Vienna; New York, Springer, 2009.

Parker, L. & Blodgett, J., *Greenhouse Gas Emissions: Perspectives on the Top 20 Emitters and Developed versus Developing Nations*, Congressional Research Service, 2010. Available at <www.crs.gov RL32721>. CSR Report.

Peel, J., Godden, L. & Keenan, R.J., 'Climate Change Law in an Era of Multi-Level Governance', 1(2) *Transnational Environmental Law*, 2012. Available at <http://journals.cambridge.org/abstract_S2047102512000052>.

Pogge, T. (Ed.), *Freedom from Poverty as a Human Right. Who Owes What to the Very Poor*, Oxford University Press, 2007.

Pogge, T.W., 'Eradicating Systematic Poverty: Brief for a Global Resources Dividend', 2(1) *Journal of Human Development*, 2001.

Posner, E.A. & Sunstein, C.R., 'Climate Change Justice', 96 *The Georgetown Law Review*, 2008, p. 1565.

Pradeeph, B. & Madhur, Y., *New Judicial Roles and Green Courts in India*. Available at <http://inece.org/conference/9/proceedings/08_Scanlon.pdf>.

Preston, B.J., *The Role of the Judiciary in Promoting Sustainable Development: The Experience of Asia and the Pacific*. Available at <www.lawlink.nsw.gov.au/agdbase7wr/_assets/lec/m420301l721754/speech_10jan2006_preston.pdf>.

Preston, B.J., *Administrative Law in an Environmental Context: An Update*, 2007. Available at <www.nsw.gov.au>.

Pring, G. & Pring C., *Greening Justice, Creating and Improving Environmental Courts and Tribunals*. The Access Initiative, 2009. Available at <www.law.pace.edu/sites/default/files/IJIEA/jciPring%20PACE-IJIEA%203-17_cropped.pdf>.

PWC, *The Future of Reporting, Carbon Disclosure Project*, 2012. Available at <www.cdproject.net>.

Rajamani, L., 'India and Climate Change: What India Wants, Needs, and Needs to Do', 8(3) *India Review*. Available at <http://dx.doi.org/10.1080/14736480903116842>, pp. 340-374.

Rajamani, L., 'The Changing Fortunes of Differential Treatment in the Evolution of International Environmental Law', 88(3) *International Affairs*, 2012, pp. 605-623.

Rajamani, L., 'Rights Based Climate Litigation in the Indian Courts: Potential, Prospects & Potential Problems', Working Paper 2013/1 (May), available at <www.cprindia.org>.

Rest, A., 'Enhanced Implementation of International Environmental Treaties by Judiciary – Access to Justice in International Environmental Law for Individuals and NGO's: Efficacious Enforcement by the Permanent Court of Arbitration', 1 *Macquarie Journal of International and Comparative Environmental Law*, 2004.

Rieter, E., *Preventing Irreparable Harm, Provisional Measures in International Human Rights Adjudication*, Antwerp, Intersentia, 2010.

Robertson, D.W., *Causation in the Third Torts Restatement: Three Arguable Mistakes*. Available at <tortssymposium.law.wfu.edu/papers/Robertson.pdf>.

Rogers, W.V.H., *Winfield & Jolowicz on Tort* (18th edn), Sweet & Maxwell, 2010.

Rosen-Zvi, I., 'Israeli Report', *in* Lord, R. *et al.* (Eds.), *Climate Change Liability.*

San José, D.G., *Environmental Protection and the European Convention on Human Rights*, Council of Europe Publishing, 2005.

Santosa, M-A., Khatarina, J. & Assegaf, R.S., 'Indonesian Report', *in* Lord, R. *et al.* (Eds.), *Climate Change Liability.*

Sappideen, C. & Vines, P (Eds.), *Fleming's The Law of Torts*, Lawbook Co, 2011.

Schrobe, M., 'Fracking Outspaces Science on Its Impact', *The Journal of the Yale School of Forestry & Environmental Studies*, 2012.

Schwartz, V.E., Kelly, K. & Partlett, D.F., *Prosser, Wade and Schwartz' Torts, Cases and Materials*, Foundation Press, 2010.

Shapo, M.S., *Tort Law and Culture*, Carolina Academic Press, 2003.

Shelton, D., *Remedies in International Law* (2nd edn), Oxford University Press, 2005.

Shelton, D. & Kiss, A., *Judicial Handbook on Environmental Law*, UNEP, 2005. Available at <www.unpei.org?PDF/.../Judial-Handbook-Environment-Law.pdf>.

Simons, K.W., 'The Hand Formula in the Draft Restatement (Third) of Torts: Encompassing Fairness as Well as Efficiency Values', 54(3) *Vanderbilt Law Review*, 2001, p. 901.

Smith, J. & Shearman, D., *Climate Change Litigation*, Presidian, 2006.

Spier, J. (Ed.), *Unification of Tort Law: Causation*, The Hague, Kluwer Law International, 2000.

Spier, J., *The Law as a Crowbar for the Eradication of Poverty*, InDret, 2007, p. 406.

Spier, J., 'The Rule of Law and Judicial Activism: Obstacles for Shaping the Law to Meet the Demands of a Civilized Society, Particularly in Relation to Climate Change', *in* Faure, M. & van der Walt, A. (Eds.), *Globalization and Private Law, The Way Forward*, Cheltenham, Northampton, Edward Elgar, 2010.

Spier, J., *Shaping the Law for Global Crises*, The Hague, Eleven International Publishing, 2012.

Stahel, W.R., 'Perception, Assessment and Management of Risks and Opportunities; Different Approaches to Dealing with Risks', 52 *Risk Management Newsletter*, November 2012.

Stein, P.L., 'Judges Active in Promoting Environmental Law Capacity Building', 33(2) *Environmental Policy and Law*, 2003.

Stephenson, S., 'Jobs, Justice, Climate: Conflicting State Obligations in the International Human Rights and Climate Change Regimes', 42 *Ottawa Law Review*, 2010-2012, p. 155.

Stern, L., 'Developing Countries Must Make Deeper Emission Cuts', *The Guardian*, 4 December 2012.

Stern, N., 'What We Risk and How We Should Cast the Economics and Ethics', *Lionel Robbins Memorial Lecture Series, Lecture 1*, 2012. Available at <www.cep.lse.ac/uk/_new/events/event.asp?id-140>.

Stern, N., 'How Can We Respond and Prosper: The New Energy-Industrial Revolution', *Lionel Robbins Memorial Lecture Series, Lecture 2*, 2012. Available at <www.cep.lse.ac/uk/_new/events/event.asp?id-140>.

Supreme Court Commissioners, *Tenth Report of the Commissioners of the Supreme Court. Permanent Shelters for Urban Homeless Populations. The National Report on Homeless for Supreme Court of India. Review of Compliance of State Governments with Supreme Court Orders up to 31 December 2011*, 196 *WR* 2001.

Takamura, Y., 'Japanese Report', *in* Lord, R. *et al.* (Eds.), *Climate Change Liability. Transnational Law and Practice*, Cambridge University Press, 2012, pp. 233.

The Economist, 'The Promised Gas Revolution Can Do the Environment More Good than Harm'. Available at <www.economist.com/node/21556249/print>.

The Royal Society, *Geoengineering the Climate, Science, Governance and Uncertainties*, 2009. Available at <http://royalsociety.org/uploadedFiles/Royal_Society_Content//policy/publications/2009/8693.pdf>.

The Royal Society and the Royal Academy of Engineering, *Shale Gas Extraction in the UK, A Review of Hydraulic Fracturing*, 2012. Available at <www.roeng.org.uk/shale>.

Trindade, F., Cane, P. & Lunney, M., *The Law of Torts in Australia* (4th edn), Oxford University Press, 2007.

Umlass, E., *Corporate Human Rights Reporting: An Analysis of Current Trends.* Available at <www.globalreporting.org/.../Human_Rights_analysis_trends.pdf>.

UNDP, *Human Rights and the Millennium Development Goals.* Available at <www.undp.org/oslocentre>.

UNEP, *Towards a Green Economy, Pathways to Sustainable Development and Poverty Eradication*, 2011. Available at <www.unep.org/greeneconomy/Portals/88/documents/ger/GER_synthesis_en.pdf>.

UNEP, *Bridging Gaps Watts New?*, 2012. Available at <www.unep.org/english/climatechange/Cop18-Publication/tabid/105542/Default.aspx>.

UNEP, *Training Manuel on International Environmental Law.* Available at <www.iucnaecl.org/en/.../unep-training-manual.html>.

US Environmental Protection Agency, *Study of the Potential Impacts of Hydraulic Fracturing on Drinking Water Resources: Progress Report*, Washington, 2012.

van Dam, C., *Zorgvuldigheidsnorm en aansprakelijkheid*, Deventer, Kluwer, 1989.

van Dam, C., *European Tort Law*, New York, Oxford University Press, 2006.

van Dijk, C., 'Civil Liability for Global Warming in the Netherlands', *in* Faure, M. & Peeters, M. (Eds.), *Climate Change Liability*, Cheltenham; Gloucester, Edward Elgar Publishing, 2011.

Vansweevelt, T. & Weyts, B., *Handboek Buitencontractueel Aansprakelijkheidsrecht*, Antwerpen, Intersentia, 2009.

Verheyen, R., *Climate Change Damage and International Law, Prevention Duties and State Responsibility*, Leiden; Boston, Martinus Nijhoff Publishers, 2005.

Viney, G. & Jourdain, P., *Traité de droit civil, Les conditions de la responsabilité* (3rd edn), L.G.D.J., 2006.

Von Bar, C., *Principles of European Law, Non-Contractual Liability Arising out of Damage Caused to Another (PEL Liab. Dam.)*, Oxford University Press, 2009.

WBGU, *Solving the Climate Dilemma: The Budget Approach*. Available at <www.wbgu.de>.

Weinbaum, A., 'Unjust Enrichment: An Alternative to Tort Law and Human Rights in the Climate Change Context', 20(2) *Pacific Rim Law & Policy Journal*, 2011, p. 429.

Wibisana, M.R.A.G., *Law and Economic Analysis of the Precautionary Principle*, Maastricht, Datawyse Maastricht, 2008.

Winiger, B. *et al.* (Eds.), *Digest of European Tort Law, Volume 1: Essential Cases on Natural Causation*, Wien; NewYork, Springer, 2007.

Wiskel, B., *The Emperor's New Climate, Debunking the Myths of Global Warming*, Evergreen Environmental Company, 2006.

Wood, M.C., *Atmospheric Trust Litigation*. Available at <http://law.oregon.edu/facultydocs/mwood/atmospherictrustlitigation.pdf>.

World Bank, *Turn Down the Heat, Why a 4°C Warmer World Must Be Avoided*, 2012. Available at <www.worldbank.org>.

World Bank Group, *Development and Climate Change. A Strategic Framework for the World Bank Group*. Completion Report FY09-11. Available at <www.worldbank.org/climatechange>.

World Bank, *Convenient Solutions to an Inconvenient Truth*, Report No. 51838, 2009.

Wright, R.W., 'Once More into the Bramble Bush: Duty, Causal Contribution, and the Extent of Legal Responsibility', 54(3) *Vanderbilt Law Review*, 2004.

Zander, J., *Different Kinds of Precaution. A Comparative Analysis of the Application of the Precautionary Principle in Five Different Legal Orders*, Maastricht, 2009.

2 Injunctive Relief against Climate Change

Ulrich Magnus

2.1 Introduction

Island states, prominently in the Pacific,[1] and countries with flat coast zones[2] are particularly exposed to, and affected by, the consequences of climate change. Global warming leads to a rising sea level,[3] which threatens those states, partly in their very existence. They face severe flooding or even the complete extinction of their land. There is little doubt that mankind at least contributes significantly to global warming and thus to these consequences.[4] A cynic would probably say that the population of the affected states has just to move to other regions of the globe. For lawyers this scenario raises, among many other issues, the question whether those states or their people could, by legal means, pursue claims against others for the loss, or rather the threatened loss, of their land. Compensation after the event would evidently be of no great avail. It is prevention that is needed. In environmental matters the most effective measures are regularly those which immediately stop activities that damage or threaten to damage the environment as well as other people. In law this means injunctive relief.[5] Do any respective legal possibilities exist for those states or their citizens? Can they institute legal proceedings for injunctive relief?

1 For instance the State of Palau, the Maldives, the Marshall Islands etc. Indeed, on 22 September 2011 Palau's President Johnson Toribiong urged the United Nations Plenary Meeting to ask the International Court of Justice for an advisory opinion on the responsibility of states for the consequences of climate change; *see* A. Zimmermann & J. Bäumler, 'Wer ist schuld', *Frankfurter Allgemeine Zeitung*, 28 June 2012, p. 8.

2 For instance Bangladesh or the Netherlands.

3 The IPCC Report of 2007 estimates a rise by 19 to 59 cm by the end of the 21st century: *see* IPCC Fourth Assessment Report: Climate Change 2007: Working Group II 6.3.2 <www.ipcc.ch/publications_and_data/ar4/wg2/en/ch6s6-3-2.html>. Other estimates even range from 50 to 190 cm: S. Rahmstorf, 'A Semi-Empirical Approach to Projecting Future Sea-Level Rise', 15 *Science*. 5810, 2007, pp. 368-370; M. Vermeer & S. Rahmstorf, 'Global Sea Level Linked to Global Temperature', *Proceedings of the National Academy of Science of the United States of America* 2009. According to the Australian Commonwealth Scientific and Industrial Research Organisation (CSIRO), the rate of rise in the last two decades was more than 50% higher than the average rate over the entire 20th century; *see* CSIRO, <www.cmar.csiro.au/sealevel/sl_hist_last_15.html>; also R.J. Nicholls & A. Cazenave, 'Sea-Level Rise and Its Impact on Coastal Zones', 328 *Science*, 5985, 2010, pp. 1517-1520.

4 *See* thereon IPCC Fourth Assessment Report: *Climate Change 2007:* Synthesis Report 2.2, <www.ipcc.ch/publications_and_data/ar4/syr/en/mains2-2.html>; *see also* the account given by J. Spier, *Shaping the Law for Global Crises*, Den Haag, Eleven International Publishing, 2012, pp. 11-25.

5 *See also* Spier, 2012, p. 194.

An injunction against activities that cause the climate change may sound rather strange: Injunctions by whom against whom? Against which activities? In which courts? Under which law? For which reasons?

When these questions are raised, the procedural and the substantive aspects of injunctive relief have to be distinguished: The procedural side comprises the question of international jurisdiction for, and enforcement of, an injunctive judgment or order; the substantive side comprises the question of which law applies and which rules the applicable law provides for. The following text examines the existing possibilities of injunctive relief, both in procedural and in substantive respect, and asks whether they suffice or whether they should be enlarged, and if so in which direction.

In discussing possible legal – generally civil law – instruments of avoiding, or at least protecting against, dangerous and irreversible climate change, this does not at all mean that such legal instruments are the only remedies. There are many others,[6] primarily of a public law nature, for instance prevention and control by administrative bodies that license and supervise industries, plan and steer the use of the environment etc. In view of the urgency to keep the change of the world climate within still controllable and bearable boundaries, all these possibilities need to be examined, often improved and efficiently pursued. In particular, the precautionary principle[7] has to be observed as a general guideline for states, legal and natural persons to avoid environmental damage. However, though this principle as such and also compensation for environmental damage[8] have been intensely discussed, legal instruments of an injunctive nature that can be used in lawsuits, have so far played a comparatively little role in the global warming discussion.[9] It is the aim of the following text to examine whether they can be used as a means of preventing environmental damage and whether they can effectively support the objective to maintain liveable conditions on the globe.

---

6 For a survey *see* the introductory article by R. Moncel & H. van Asselt, 'All Hands on Deck! Mobilizing Climate Change Action beyond the UNFCCC', 21 *Review of European Community & International Environmental Law* 2012, pp. 163 *et seq.*, and further contributions to this special issue of the journal; also D. Bodansky, *Multilateral Climate Efforts beyond the UNFCCC*, Center for Climate and Energy Solutions 2011.

7 *See*, in particular, Principle 15 Rio Declaration of 1992: "In order to protect the environment, the precautionary approach shall be widely applied by all States according to their capabilities. Where there are threats of serious or irreversible damage, lack of full scientific certainty shall not be used as a reason for postponing cost-effective measures to prevent environmental degradations"; further J. Cameron & J. Abouchar, 'The Status of the Precautionary Principle in International Law', *in* D. Freestone & E. Hey (Eds.), *The Precautionary Principle and International Law*, The Hague, Kluwer 1996; A. Trouwborst, *Evolution and Status of the Precautionary Principle in International Law*, The Hague, Kluwer Law International 2002; A. Proelß, *in* W. Graf Vitzthum (Ed.), *Völkerrecht* (5th edn), Berlin/New York, de Gruyter, 2010, pp. 465 *et seq.*

8 *See*, for instance, E.H.P. Brans, *Liability for Damage to Public Natural Resources. Standing, Damage and Damage Assessment*, The Hague, Kluwer Law International, 1996; M. Faure & A. Verheij (Eds.), *Shifts in Compensation for Environmental Damage*, Wien, New York, Springer, 2007; *see also* the liability conventions in the fields of marine oil pollution, nuclear energy etc. which provide for compensation of damage but do not grant remedies before damage has occurred.

9 But *see* Spier, 2012, pp. 194 *et seq.*

2.2 What Injunctive Relief Means

Injunctive relief means a court's decision that orders the defendant to do or, in particular, not to do a specific act – the act is more than merely paying a sum.[10] The central aim of injunctive relief is the avoidance of a threatened damage or violation of a right. By an injunction a person can be ordered to refrain from doing or continuing a certain activity endangering others, for instance emitting CO_2 beyond a defined threshold if thereby injury to others threatens. In environmental cases, injunctive relief will very often order the omission of a certain activity, though only under specific circumstances. However, as injunctive relief can also oblige the defendant to act in a certain positive way, this can be particularly relevant with respect to states, where for instance an order may oblige a state to implement or enact certain legislation against climate change. It is evident that this kind of order is especially problematic because it interferes with the state's sovereign power to legislate. If at all admissible, its preconditions will be rather restrictive.

Injunctive orders can be either provisional or final. The formalities and preconditions of both kinds of injunctions may vary considerably. Generally, the requirements for a preliminary decision are less strict than for the final judgment as far as proof and the certainty of the threatened risk are concerned. Where climate change is endangering the existence of a country, potential claimants will most likely prefer a final decision ultimately prohibiting activities that cause or add to global warming or are at least likely to cause or add thereto and thus to consequential loss of land and property. Yet, a provisional decision that forbids such activities until a final decision may also be of considerable interest.

Further, injunctive relief may be asked for either where no damage has yet occurred but is threatening or where some damage has already occurred and further future damage is impending. In both situations, injunctive relief requires a specific standard of probability of threatened future damage. Generally, this standard will be stricter for the first of the two situations: where damage is expected for the first time, its entry must be rather likely before injunctive relief is granted. Where an activity has already caused damage or even continues to cause it, then the likelihood of future damage can be much easier to assume.

Generally, injunctive relief and also the application for such relief must very precisely circumscribe the act that is requested to be done or to be omitted. In particular, in case of omissions it can be necessary to list a number of clearly defined acts that shall be prohibited.

Specific problems are linked with the execution of injunctive orders. If the defendant does not comply with the order, it is not always possible to enforce the order in kind, or at least to enforce it immediately. Therefore, it is rather usual to combine the order with the threat of a drastic money penalty for non-compliance. The enforcement of a money claim is regularly less difficult than that of a specific act or omission. On the other

10 *See* for instance E. Jowitt, *Dictionary of English Law* (2nd edn), London, Sweet and Maxwell, 1977/1985, p. 975.

hand, the enforcement of the money claim does not achieve the actual aim of avoiding – environmental – damage.

2.3 Hierarchy of Sources

Although it may seem to be a certain deviation, it is helpful for our subject to start with the various instruments and their hierarchy – treaties, conventions, acts, statutes of supranational, international and national law that can play a role here. It is a preliminary yet important question whether and in which order all these sources are to be taken into account. Although it might be astonishing to find a generally accepted answer, there is wide **unanimity on the abstract order of the mentioned sources, though details are still disputed** and certain exceptions may apply.[11] The practical consequence of the order is that the law on the higher level of hierarchy regularly prevails over any inconsistent law on a lower level and can invalidate it.

At the top rank are certain – 'absolutely' – mandatory rules of international (public) customary law or law of nations (parts of the so-called *jus cogens*). These are nondispositive rules accepted by all or almost all nations as binding and indispensable so that each state is obliged to observe them.[12] They are only few, but they are even respected by the nation states' constitutions, which in principle are the supreme law of the land.[13] According to a widely shared view, the most prominent part of this international customary law are the elementary human rights[14] as first laid down in the UN's Universal Declaration of Human Rights of 1948 and later further specified and refined by other treaties and conventions, in particular by the International Covenant on Civil and Political Rights and the International Covenant on Economic, Social and Cultural Rights, both of 1966. They may include a right to healthy water and air;[15] however, it is still unclear and disputed precisely which human rights and to which exact extent they form part of this 'absolutely'

11 It is, however, often stated that there is no hierarchy among the sources of the law of nations, although mandatory norms of customary international public law prevail over all other sources, in particular over international treaties: *see* W. Graf Vitzthum, *in* W. Graf Vitzthum (Ed.), *Völkerrecht*, Berlin/New York, de Gruyter, 2010 (5th edn), p. 69; T. Stein & C. von Buttlar, *Völkerrecht*, Cologne, Carl Heymanns (13th edn), 2012, para. 147.

12 *See* Arts. 53 and 64 Vienna Convention on the Law of Treaties. An example is the immunity that states enjoy in foreign courts as far as their acts *jure imperii* are concerned; *see* recently ICJ, judgment of 3 February 2012 (*Germany v. Italy*) <www.icj-cij/docket/files/143/16883.pdf>, paras 59 *et seq. See also* I. Brownlie, *Principles of Public International Law*, Oxford, Oxford University Press (6th edn), 2003, pp. 335 *et seq.*; K. Hailbronner & M. Kau, *in* Graf Vitzthum 2010, p. 181; Stein & von Buttlar, 2012, paras 713 *et seq.*

13 An example is the German Constitution (*Grundgesetz* – GG); it acknowledges the peremptory character of the general rules of international customary law (Art. 25 GG) and prohibits any change of the constitution that would abolish the essential core of the human rights (Art. 79[3] GG).

14 Hailbronner & Kau, 2010, p. 181; Stein & Von Buttlar, 2012, paras 713 *et seq.*

15 In this sense European Court of Human Rights, EuGRZ 1995, 530 (*López Ostra v. Spain*).

binding international customary law. A further such absolutely mandatory rule is probably the commandment that states must not permit activities on their territory that cause or threaten to cause serious damage in other states.[16] It would be unacceptable if a state by its own constitution and legislation could dispense itself from this obligation. Again, the precise limits of the rule are still vague. Other mandatory rules of international law that even a country's constitution cannot validly abolish, but that do not matter here, may be the ban on slavery or aggressive wars.

On a second level, generally, the national constitutional law encounters. It prevails even over those rules of the international *jus cogens* that are not absolutely binding. This is mainly owed to the maxim of sovereignty of the national state. The single state if recognized by other states is regarded as autarkic to enact its own law. However, constitutions used to be reluctant and vague with respect to precise provisions on the protection of the environment.[17] Often they formulate mere objectives of the state without granting specific rights to affected persons.[18]

A third level can but need not exist. It is supranational law such as the law of the European Union. As in the case of the EU, the national states may transfer parts of their legislative power onto a supranational institution. Nonetheless, they maintain their general constitutional power ('competence-competence'). The supranational law defines its effect with respect to the national law of the involved states. In the case of the EU, if not otherwise provided for, the mandatory supranational law takes precedence over the internal law of the Member States[19] and, in principle, even over their international commitments in those fields where the national state has transferred its sovereign power onto the supranational institution.[20]

The next hierarchic level is generally represented by the bi- or multilateral treaties or conventions that a state has ratified. By the respective procedure of ratification these instruments become part of the national law. Nonetheless, they regularly prevail over any deviating national law. The reason is that the enacting state normally wants to fulfil its

16 *See* ICJ Reports 1949, 4 (22) (*United Kingdom v. Albania, Corfu-Channel case*); ICJ Report 2010, p. 14 (*Argentina v. Uruguay, Pulp Mills case*) and already the *Trail Smelter Arbitration*, RIAA III 1938, 1965; *see also* Principle 21 of the Stockholm Declaration of 1972 (ILM 11 [1972] 1416) and Principle 2 of the Rio Declaration of 1992 (ILM 31 [1992] 876); both Declarations are not binding but formulate universally accepted principles.

17 *See* for instance Art. 20a German GG: "The State protects in responsibility for future generations the natural foundations of life and the animals [. . .]" or Art. 48a Indian Constitution: "The State shall endeavour to protect and improve the environment and to safeguard the forests and wild life of the country."

18 *See* the examples in the preceding footnote.

19 In the case of the EU, the effect of EU law depends on the nature of the legislative instrument (primary law or secondary law [directive or regulation]) and on the precise content of the respective norm; *see* in detail R. Bieber, A. Epiney & M. Haag, *Die Europäische Union* (10th edn), Baden-Baden, Nomos, 2013, pp. 199 *et seq.*

20 *See* indirectly Art. 351 TFEU (the Member States must do their best to adapt international treaties which they concluded before they joined the EU if those treaties are incompatible with EU law; after the accession to the EU they have no longer the competence to conclude international treaties in fields where the EU is – exclusively – competent); for details *see e.g.*, Bieber *et al.*, 2013, pp. 619 *et seq.*; T. Oppermann, C.D. Classen & M. Nettesheim, *Europarecht*, München, Beck (4th edn), 2009, pp. 669 *et seq.*

international obligations that it has undertaken in adopting the international instrument. Although the national state is free to neglect and refuse these obligations, it must generally be presumed that the state intends to observe them.[21] Therefore if the national legislator does not explicitly declare the intention to deviate from a ratified international treaty, the provisions of that treaty prevail over any inconsistent national law even if it is of later origin. In the environmental field, many international treaties exist.[22] However, thus far they appear not to provide for means of injunctive relief.

National law (except national constitutional law) ranges on the next hierarchic level. Although on all levels rules of equal rank may collide, this most likely happens on the national level. Generally, two rules are adduced to settle this conflict, actually on all levels: the more special rule derogates the more general, and the younger provision derogates the older.[23] National law provides for rules on injunctive relief as a general procedural remedy, though rarely specifically in the environmental context.

At the bottom of the hierarchy of norms, we find the so-called soft law. It comprises international sets of principles such as the UNIDROIT Principles of International Commercial Contracts or the Principles of European Contract Law as well as the INCOTERMS or like instruments. In the environmental context, the Stockholm Declaration of 1972 constitutes such soft law.[24] All these instruments are not binding by themselves but can be agreed on by the parties. Nonetheless, by their existence they contribute to the development of internationally accepted rules in that they evidence a widely recognized legal standard.

One could ask how private law agreements of parties fit into this ladder of hierarchy. It is the prevailing view that party autonomy stands under the reign of the law and is admitted only within the limits the applicable law sets on each level. Private contracts as such cannot oust the otherwise applicable law.[25]

2.4 Who Should Decide?

It is questionable whether it has to be courts that should decide on injunctive relief for environmental issues. Alternatively, one could think of arbitration or mediation institutions for this task. Also administrative bodies could be well-suited institutions to effectively prohibit activities that contribute to global warming and its consequences. Or,

21 For a comparative survey *see* P. Kunig, *in* Graf Vitzthum, 2010, pp. 96 *et seq.*

22 Particularly relevant for the climate change: UN Framework Convention on Climate Change of 9 May 1992 and its Kyoto Protocol of 10 December 1997; for a list of international environmental treaties *see* A. Proelß, *in* Graf Vitzthum 2010, pp. 391 *et seq.*

23 W. Graf Vitzthum, *in* Graf Vitzthum, 2010, p. 69; Stein & Von Buttlar, 2012, para. 154.

24 Declaration of the UN Conference on the Human Environment, ILM 11 (1972) 1416.

25 This is the clearly prevailing view in private international law; *see* U. Magnus, *in* J. von Staudinger, *Kommentar zum Bürgerlichen Gesetzbuch mit Einführungsgesetz und Nebengesetzen*, Berlin, Sellier, de Gruyter, 2011, Art. 3 Rom I-VO, paras 40 *et seq.* with numerous references.

political bodies could be mandated with the task. Though all these alternatives may come into account and should be considered, in the end a clear decision is needed that can be executed in normal execution procedures. This generally requires a conclusive court decision. An arbitration award or the decision of a mediator may still be attacked before a court of law.[26] True, certain reasons must be present for such an attack; nevertheless, it seems preferable then to directly involve a court of law. The same reasoning militates against administrative bodies as the only agencies dealing with environmental issues. Their decision will regularly be subject to judicial review before a court. It is then quicker and more efficient to involve the court directly and immediately, particularly if the court is specialized in environmental issues.[27] Political bodies are probably least apt to deliver a final and executable decision on a concrete dispute. The different political views generally represented in political bodies do not facilitate the finding of a neutral decision that is influenced solely by environmental aspects. To determine a dispute and order a specific remedy such as injunctive relief is the essential task of the judiciary.

2.5 The Problem of State Immunity

A first question when dealing with lawsuits is always whether or not a defendant is immune against all kinds of legal proceedings. State immunity is a general rule of international customary law, which saves states or their official representatives from adjudication by the courts of other states. The exemption from foreign jurisdiction is confined to cases where a state or its personnel acts in its capacity as state or state official. If states act for public purposes and through means of public law in the exercise of their sovereign public powers (*acta iure imperii*), they can rely on their immunity in foreign courts.[28] This is an acknowledged principle of (public) international customary law and has become part of the law of nations.[29] Only where a claim against a state results from the state's commercial activities – where the state acts like a private natural or legal person (*acta iure gestionis*) – no immunity in foreign courts is granted.[30] This means that acts of states in pursuit of their public powers can be attacked only in the courts of that state and to the extent foreseen

26 *See* Art. 34 UNCITRAL Model Law on International Commercial Arbitration of 1985, which has been adopted by many countries.

27 There are countries that have established specialized environmental courts. An example is India with its (Federal) Green Tribunal; *see* G. Pring & C. Pring, *Greening Justice – Creating and Improving Environmental Courts and Tribunals*, 2009, pp. 9 *et seq.* and further below.

28 *See* most recently ICJ, judgment of 3 February 2012 (*Germany v. Italy*) <www.icj-cij/docket/files/143/16883.pdf>, paras 59 *et seq*; further: Brownlie, 2003, pp. 335 *et seq.*; Hailbronner & Kau, *in* Graf Vitzthum 2010, p. 181; Stein & Buttlar, 2012, pp. 270 *et seq.*; particularly to the ICJ judgment: B. Hess, 'Staatenimmunität und ius cogens im geltenden Völkerrecht: Der Internationale Gerichtshof zeigt die Grenzen auf', 32 *Praxis des Internationalen Privat- und Verfahrensrechts*, 2012, pp. 201, 206.

29 *See* ICJ ibid, n. 28.

30 ICJ ibid, n. 28; Hailbronner & Kau, *in* Graf Vitzthum, 2010.

there. In proceedings against states concerning climate change most likely the omission of appropriate legislation or lack of correct implementation will be at stake. This either legislative or administrative activity or often its omission hardly qualifies other than as *actus iure imperii*; to legislate or not to legislate is the very exercise of the state's sovereign power; it is state activity par excellence. The same holds true for the exercise of a state's administrative powers. As a consequence a state enjoys immunity in foreign courts with respect to its legislative and administrative activity or inactivity.
In the environmental context, only the immunity of states (not of persons) is of interest and relevance. Immunity here includes jurisdictional immunity, immunity against any enforcement of a judgment and in a sense also the question whether a state that has accepted the jurisdiction of a foreign court can be ordered to do or not do acts of state power, *acta jure imperii*. Further, it has to be examined separately whether a state can invoke the immunity defence when sued before the International Court of Justice (ICJ), a supranational or a national court, and non-compliance with standards of international environmental law, be it treaty law or customary law, is alleged. Any injunctive order against the defendant state would condemn it to act either as legislator or on the administrative level, necessarily and regularly involving *acta jure imperii*.

2.5.1 State Immunity before the International Court of Justice

Before the ICJ, a defendant state can raise the defence of lacking jurisdiction in each case where the state has not validly submitted to the jurisdiction of the ICJ.[31] In this sense, a state that has not submitted is immune before the ICJ. In consequence, states can avoid any adjudication by the ICJ if they do not accept the Court's jurisdiction.
But even where a state has submitted to the Court and enjoys no jurisdictional immunity before the ICJ, it may be questioned whether the state can still raise a defence – so to speak of sovereign immunity – against being adjudicated to *acta jure imperii*. In a case concerning execution into Germany's property as compensation for German war crimes in World War II, the ICJ explicitly rejected the argument that international *jus cogens* rules – in that case human rights – can overthrow the doctrine of state immunity for *acta jure imperii*.[32] However, in that case the ICJ had merely to decide whether national states and their courts were bound by the doctrine. Although the ICJ formulated its statement in rather general terms, the Court has not yet expressed it as a rule to be observed either by itself or in the context of international environmental disputes.

31 *See* ICJ, *Argentina v. Uruguay (Pulp Mills)* paras 48 *et seq.*; by *argumentum e contrario* this also follows from Arts. 7-9 UN Convention on Jurisdictional Immunity of States and their Property of 2004 and Art. 2 European Convention on State Immunity of 1972, which express that a state cannot invoke its immunity where it has submitted to the jurisdiction of another state.

32 *See* ICJ, 3 February 2012 (*Germany v. Italy*; Greece intervening) (German war crimes in Italy and Greece did not justify an exception to the general principle of state immunity).

On the contrary, in the *Pulp Mills* case[33] – a "case on international environmental law of an exemplary nature"[34] – the Court did not raise the immunity issue in this sovereign sense although in that case Argentina alleged that Uruguay had violated international obligations by not prohibiting the erection of private pulp mills at the riverside of the river Uruguay. Uruguay could have complied with its obligation only through legislative or administrative action, thus acting in its public capacity. The ICJ denied a breach of substantive international environmental duties, but held that Uruguay had violated a procedural duty to inform Argentina of the project in a correct manner. Even this duty Uruguay could have performed only by an act of public power, not by an *actus jure gestionis*. From the *Pulp Mills* decision it has to be inferred that a state after having accepted the ICJ's jurisdiction cannot further invoke its sovereignty as defence against a compensation claim, where the non-compliance with binding international obligations has caused damage in another country even if avoiding that damage would have demanded the state's acting as state. A supporting argument can be derived from Article 12 UN Convention on Jurisdictional Immunity of States and their Property of 2004 and Article 11 European Convention on State Immunity of 1972. Both provisions deny (jurisdictional) state immunity before national courts in case of compensation for bodily harm or property damage where the damage occurred in the forum state and the author then was there. It does not matter whether compensation has to be made by an *actus jure imperii*. From *Pulp Mills* it has to be taken that in proceedings before the ICJ the sovereignty defence is not available in cases of non-compliance with binding international environmental obligations even if a state can regularly perform this obligation only by acting per *acta jure imperii*. The reasoning behind this would be the consideration that the compliance with international environmental obligations, which almost always affect other states and whose neglect often causes damage there, can only be achieved if the state uses its legislative and/or administrative means and its state power. Here to permit a defence of sovereign immunity would render the obligations more or less futile and deprive the states that suffer from the non-compliance of an effective remedy. Such an extension of state immunity in international environmental disputes would considerably exacerbate the situation of those affected by any violation of international environmental obligations.

2.5.2 State Immunity before Regional and National Courts

Jurisdictional state immunity and immunity in the enforcement of judgments against states is to be respected in regional and national courts as well.[35] Where the European

33 ICJ, *Argentina v. Uruguay* (*Pulp Mills*).

34 ICJ, *Argentina v. Uruguay* (*Pulp Mills*), Joint Dissenting Opinion Al-Khasawneh and Simma, para. 3.

35 *See* thereto for instance German Bundesverfassungsgericht (Federal Constitutional Court) BVerfGE 16, 27 *et seq.*; 46, 342 *et seq.*; 64, 1 *et seq.*; H. Schack, *Internationales Zivilverfahrensrecht* (5th edn), Munich, Beck, 2010, paras 145 *et seq.*; Hailbronner & Kau, *in* Graf Vitzthum, 2010, pp. 180 *et seq.*

Convention on State Immunity of 1972 is in force,[36] the – jurisdictional – immunity rules of this Convention apply. However, they mirror more or less the standard of the international customary law. As long as states act in their state power for state purposes, they are in principle immune before regional and national courts.
On the regional level the European example must suffice. Neither the Brussels I Regulation nor the Lugano Convention 2007 supersedes international customary law. If under international law a state enjoys immunity in foreign courts, the Regulation and the Convention respect this.[37] Thus, under the Regulation and the Convention, injunctions against legislative or administrative inactivity of states cannot be claimed. Such claims fall outside the scope of the Regulation[38] and the Convention.
On the contrary, insofar as states are involved in commercial activities in the same way as private natural or legal persons, the Regulation's and the Convention's normal jurisdiction rules apply. For instance, the production and sale of energy by state-owned entities has to be, and has been, regarded as civil and commercial activity.[39] To such cases the rules applicable to private defendants apply.
On the national level we find the same picture. The national courts, if not anyway bound by the European Convention on State Immunity, have to, and will, observe international customary law and thus accept the jurisdictional state immunity defence.[40] In consequence, states when acting in their capacity as states cannot be sued before national courts of other countries unless they voluntarily submit to the respective foreign jurisdiction.

2.6 Jurisdiction and Enforcement of Judgments

If courts are regarded as the right forum for injunctions against activities causing climate change, they can decide only if they possess jurisdiction. Furthermore, even if a court has jurisdiction and renders an injunctive order, this would be of no avail to a potential claimant unless the order could be enforced in a country where the defendant is acting or at least has assets. Thus, jurisdiction is a necessary though not always sufficient requirement for effective legal protection of claimants. Besides jurisdiction, also enforcement of

36 Only 8 EU Member States are Contracting State of the Convention.
37 *See* European Court of Justice [2007] ECR I-1519 (C-292/05, *Lechouritou v. Germany*); *see* thereto R.Geimer, 'Los Desastres de la Guerra und das Brüssel I-System', 28 *Praxis des Internationalen Privat- und Prozessrechts*, 2008, pp. 225-227; M. Stürner, 'Staatenimmunität und Brüssel I-Verordnung', 28 *Praxis des Internationalen Privat- und Prozessrechts*, 2008, p. 204 .
38 European Court of Justice [2007] ECR I-1519 (C-292/05, *Lechouritou v. Germany*).
39 *See*, for instance, Amtsgericht Bonn, *Neue Juristische Wochenschrift* 1988, p. 362; Landgericht Bonn, *Neue Juristische Wochenschrift*, 1989, p. 1225 (regarding the energy production by the Soviet nuclear reactor in Tshernobyl as civil and commercial activity).
40 *See*, for example, for Austria: OGH (Supreme Court) Juristische Blätter 1986, pp. 733 *et seq.*; for Germany: German Bundesverfassungsgericht (Federal Constitutional Court) BVerfGE 16, 27 *et seq.*; 46, 342 *et seq.*; 64, 1 *et seq.*; for the United Kingdom: *Alcom Ltd. v. Republic of Colombia* [1984] 2 All E.R. 6 *et seq.* (H.L.).

judicial measures must be secured. Therefore both aspects have to be explored. First, is there, and if so where, a jurisdictional basis to institute proceedings against defendants who may be held responsible to have contributed to climate change? Secondly, can a judgment be enforced in other countries?
Jurisdiction here means, in the first place, international jurisdiction, namely whether and when a court is competent to hear international cases. In the second place, jurisdiction concerns the question of what kinds of disputes *ratione materiae* can be brought before a court that is internationally competent. Enforcement means the execution of a judgment either in kind or by way of money sanctions. For injunctive relief, execution in kind would be particularly effective because it would directly achieve the aim of the injunction.
Always linked to the aspect of jurisdiction is the question of who can be sued. Possible defendants can be either states or private natural or legal persons. As will become apparent below, the jurisdictional requirements vary to a certain extent depending on whether states or private persons are sued.

2.6 1 International Level

For global challenges as existential as climate change, a globally competent court with specialized knowledge on environmental and climate change issues and effective means of enforcement would probably be the intuitive and natural answer.[41] And in certain fields, *e.g.* criminal law,[42] war crimes,[43] certain maritime matters[44] or on trade disputes within the WTO,[45] there already exist specialized 'world courts'. There, the international community has been able to establish global judicial reactions and measures. Partly, these courts are competent even without the necessity that the state or the party against whom action is taken must have submitted to the jurisdiction of the court.[46] However, for environmental

41 Such a court has been asked for and its establishment has been discussed for almost two decades: *see*, for instance, A. Postiglione, *The Global Environmental Crisis: The Need for an International Court of the Environment*, 1996; A. Rest, 'The Indispensability of an International Environmental Court', 7 *Review of European Community and International Environmental Law*, 1998, p. 63; E. Hey, *Reflections on an International Environmental Court*, Martinus Nijhoff Publishers, 2000; T. Stephens, 'Multiple International Courts and the "Fragmentation" of International Law', 25 *Australian Yearbook of International Law*, 2006, p. 227; T. Stephens, *International Courts and Environmental Protection*, Cambridge, Cambridge University Press, 2009; A. Akhtarkhavari, *Global Government of the Environment: Environmental Principles and Change in International Law and Politics*, Cheltenham, Edward Elger Publishing, 2010; for a detailed discussion *see* below.

42 The International Criminal Court in The Hague.

43 The International Criminal Tribunal for the former Yugoslavia in The Hague.

44 The International Tribunal of the Law of the Sea in Hamburg.

45 *See* the Panels of the WTO.

46 This is, *e.g.*, the case under Art. 13 of the Statute of the International Criminal Court: the Court can be seized by a reference of the United Nations Security Council even if the situation has no further connection with a State that is party to the Rome Statute of the International Criminal Court.

challenges no specialized international court exists yet. Do other international possibilities for court proceedings exist and do they suffice?

2.6.1.1 Limited Jurisdiction of the ICJ

The first question is whether, and if so how far, the ICJ in The Hague is competent for proceedings in which injunctive relief against climate change would be claimed. According to the Court's Statute any Member State of the United Nations can approach the ICJ. However, the Court is competent for disputes between states only;[47] proceedings by or against private persons or entities cannot be instituted there.
Furthermore, the defendant state must have submitted to the jurisdiction of the ICJ in one way or the other as detailed by Articles 35-37 Statute of the ICJ. Otherwise, the ICJ has no jurisdiction. However, only one-third of the UN Member States (67 out of 193) have ratified the ICJ Statute and have thus generally submitted to the jurisdiction of the Court by express declaration; some of these states have even done so under certain (permitted) reservations only.[48] Among the states entirely abstaining from the ICJ's jurisdiction are, for instance, the USA, China, Russia or Brazil, which are particularly important for the world's climate. They cannot be sued before the ICJ unless they would submit or specifically agree to such proceedings, which seems rather unlikely.
Thus, as far as jurisdiction is concerned, a UN Member State could seize the ICJ with a claim for injunctive relief against a number of other states, but not against all.[49] The greater number of states and, in particular, climate-important states cannot be brought before the ICJ nor can private persons approach the Court. In sum, the Court's position as an international 'environmental world court' is very limited.

2.6.1.2 No Specialization in Environmental Matters

A further point of concern is the Court's lack of special experience and knowledge in environmental matters. Although the Court instituted a special Environmental Chamber in 1993,[50] this Chamber was never approached by parties, and the Court abandoned the practice of electing members of the special Chamber in 2006.[51] This Chamber no longer exists (though it could be revived at any time). The lack of use of this special Chamber may have several reasons: it may be due to the fact that it was composed of judges of the ICJ and not of persons of special knowledge and experience in international environmental law.

47 *See* Art. 34(1) ICJ Statute.

48 *See* the list of ratifications and declarations <www.icj-cij.org/jurisdiction/index.php?p1=5&p2=1&p3=3>; further M., *in* Graf Vitzthum, 2010, p. 625.

49 Art. 41 ICJ Statute empowers the Court to grant provisional measures for the preservation of a party's rights if the Court considers "that circumstances so require."

50 ICJ press release 93/20 of 19 July 1993. The Environmental Chamber was established under Art. 26(1) Statute of the ICJ. Seven ICJ judges formed the special Chamber.

51 *See* ICJ website, <www.icj-cij.org/court/index.php?p101&p2=4>.

Moreover, states may have wished to receive a decision from the full Court (with at least nine judges) instead of the Environmental Chamber with at best seven judges. Also, few of the disputes brought before the ICJ had their sole focus on environmental questions. Thus far, there is therefore only limited case law of the ICJ on environmental matters[52] and the ICJ's contribution to international environmental law has certainly not been overstated when it was described as "relatively modest."[53]

2.6.1.3 Enforcement of ICJ Decisions

A rather huge problem is the enforcement of judgments of the ICJ. According to Article 59 Statute of the ICJ, the Court's judgments are binding between the parties. The Statute does, however, not provide for any enforcement procedure. According to Article 94(1) UN Charter, all UN Member States have accepted the obligation to observe and follow judgments of the ICJ if bound by the judgment.[54] If states do not voluntarily comply with this obligation, according to Article 94(2) UN Charter, the Security Council of the UN may "make recommendations or decide upon measures to be taken to give effect to the judgment" of the ICJ.[55] This possibility has been used only on very few occasions, in particular in the *Anglo-Iranian Oil Company* case of 1951, the *Nicaragua* case of 1986 and the *Bosnia* case of 1993.[56] It is the prevailing view that the Security Council has discretion whether or not to enforce a judgment of the ICJ, and it is further clear that the decision of the Security Council depends on political rather than legal or special environmental considerations. Although the mechanism of Article 94(2) UN Charter could be used for the enforcement of ICJ judgments on environmental matters, it is not likely that it would often be so used. In particular, where the state that does not obey an ICJ judgment is a member or even a permanent member of the Security Council, it is more than doubtful whether the Council would and could take any action against that state.[57]

A further problem is the kind of enforcement provided that the Security Council decides to take action. As already mentioned, the most effective measures for the protection of the

52 The most recent case is ICJ, *Argentine v. Uruguay (Pulp Mills)*, <www.icj-cij.org/docket/files/135/15877.pdf>.

53 *See* J.E. Vinuales, *The Contribution of the International Court of Justice to the Development of International Environmental Law: A Contemporary Assessment*, 2008, available at <http://works.bepress.com/jorge_vinuales/2>.

54 Art. 94(1) UN Charter reads as follows: "(1) Each member of the United Nations undertakes to comply with the decision of the International Court of Justice in any case to which it is a party."

55 Art. 94(2) UN Charter reads: "(2) If any party to a case fails to perform the obligations incumbent upon it under a judgment rendered by the Court, the other party may have recourse to the Security Council, which may, if it deems necessary, make recommendations or decide upon measures to be taken to give effect to the judgment."

56 *See* thereto A. Tanzi, 'Problems of Enforcement of Decisions of the International Court of Justice and the Law of the United Nations', 6 *EJIL*, 1995, pp. 539-572.

57 The permanent members of the Security Council (China, France, Great Britain, Russia, USA) have a veto right by which they can avoid any resolution of the Council. Whether a Council Member and, in particular, a permanent Council Member that is party to the dispute on which the judgment was rendered must abstain from voting in the Security Council is questionable.

environment are regularly those that immediately stop activities that damage or threaten to damage the environment and others. Generally, this can only be achieved by the internal legal mechanisms under national law. The Security Council has no means to directly stop such activities or to directly enact or enforce internal legislation that complies with the respective ICJ judgment. The Council can only take political measures that exert pressure on the non-complying state.

2.6.1.4 Other International Instruments

Also other international instruments with a global approach provide for a jurisdictional basis for injunctive claims (although they do not deal directly with injunctions). In particular, the UN Framework Convention on Climate Change of 1992 introduced the possibility that states that are parties to that Convention may declare that they submit any disputes concerning the interpretation and application of the Convention to the ICJ.[58] Such submission would also apply to instruments related to the Framework Convention, as for instance the Kyoto Protocol.[59] However, there appear to be no respective declarations thus far.[60] The jurisdiction provision is thus practically useless. Again, this jurisdiction, if applicable, would not be open to private claimants and defendants.
Further, except for jurisdiction agreements,[61] no global jurisdiction regime exists, and even the globally intended Hague Convention on Choice of Court Agreements of 2005 is not yet in force. Moreover, it is most unlikely that jurisdiction agreements may play a role in situations discussed here.

2.6.1.5 Private Initiatives to Establish an International Environmental Court

There are some interesting private international initiatives to institute an International Environmental Court (IEC). Thus, in 1994 an international group of private persons has established an International Court of Environmental Arbitration and Conciliation in Mexico.[62] It is a tribunal that acts in international environmental disputes if both parties agree either on arbitration or on conciliation. There is no evidence of an extended practice of this tribunal that entirely depends on the voluntary submission of the parties. As mentioned above, an arbitration award can always be attacked for certain reasons in national state courts.
A further initiative in the same direction is undertaken by the International Court for the Environment Coalition, a private group of mainly English practitioners. They propagate the institution of a global environmental court and have thus far prepared a Draft Statute for an International Court for the Environment (ICE), which can be found on the Internet.[63]

58 Art. 14(2)(a) UN Framework Convention on Climate Change of 1992.
59 *See* Art. 19 Kyoto Protocol.
60 The status of ratifications on the UN website, however, records no declarations of submission to the ICJ's jurisdiction.
61 Hague Convention on Choice of Court Agreements of 30 June 2005.
62 Its web address is <http://iceac.sarenet.es/>.
63 *See* <http://icecoalition.com/wp-content/uploads/2012/09/Draft-Constitution1.pdf>.

These private initiatives have not yet gained sufficient political momentum that would be necessary to prompt the UN to act.

2.6.2 Regional Jurisdiction and Enforcement Regimes

While neither a globally competent international court for problems of climate change exists nor an effective global jurisdictional scheme is in place yet, there exist, however, instruments on a regional level that effectively regulate the international jurisdiction of national courts. The most prominent example is the already mentioned Brussels I Regulation of the European Union, with its jurisdiction rules, even for injunctive measures.[64] The Brussels I regime is extended by the Lugano Convention[65] to three further states, namely Iceland, Norway and Switzerland. The Brussels I Regulation and the revised Lugano Convention of 2007 are almost entirely identical. Moreover, the decisions of the European Court of Justice on the Brussels I Regulation are also leading authorities for the courts of the Lugano countries.[66]

Some other regional integration movements, such as the Organisation pour l'Harmonisation en Afrique du Droit des Affaires (OHADA), have also enacted – though only few – separate jurisdiction rules, but have thus far not installed a full and comprehensive jurisdiction scheme.[67] The Brussels/Lugano regime therefore deserves a closer look. Its rules generally replace the national jurisdiction rules of the Member States.

2.6.2.1 Jurisdiction with Respect to Natural or Legal Persons

For the purpose of determining the international jurisdiction a distinction has to be drawn between cases where private (natural or legal) persons are sued or where states are sued. In the latter case the problem of immunity arises. It has already been discussed *supra*.

Final Injunctive Decisions

The Brussels I Regulation and the Lugano Convention cover merely civil and commercial matters. However, activities of private persons that potentially add to global warming and that claimants might intend to prohibit by an injunction because of threatened damage to them would generally fall within the scope of the Regulation/Convention.[68]

---

64 The EU has very recently revised the Brussels I Regulation. The new version will enter into force on 10 January 2015. The only modification that is of interest here is the abolition of the exequatur procedure within the EU.

65 Convention on Jurisdiction and the Recognition and Enforcement of judgments in Civil and Commercial Matters (originally of 1988; revised in 2007).

66 According to Art. 1(1) Protocol II to the Lugano Convention 2007, the relevant case law of the ECJ must be given "due account".

67 *See*, in particular, Arts. 3 and 20 of the Act on "Organisation des procedures simplifiées de recouvrement et des voies d'exécution". The Act entered into force on 10 July 1998.

68 *See*, in particular, Art. 5 no. 3 Brussels I Regulation/Lugano Convention: jurisdiction of "the courts in the place where the harmful event occurred *or may occur*"; for further discussion *see* below.

The general jurisdictional basis of the Regulation/Convention is the defendant's domicile.[69] If the domicile is located in a Member State, the claimant can sue there. This rule follows the worldwide accepted basic maxim *actor sequitur forum rei*: the claimant can and generally must sue the defendant at its place.

The domicile of natural persons is to be determined according to the law of the country of the alleged domicile.[70] Companies and legal persons on the contrary have their domicile either at their statutory seat, at their central administration or at their principal place of business.[71] Where claims arise "out of the operations of a branch, agency or establishment" also the courts at the place of this branch etc. are competent to decide.[72]

Under the general rule even a claimant from outside the EU can sue a defendant seated within the EU before the courts of that EU country, no matter where the event giving rise to the claim had happened.[73] The same is true for the Lugano Convention.[74] In principle, claimants from all over the world could therefore sue those enterprises in the EU that are seated here and allegedly add to global warming and endanger claimants or their countries. Whether a lawsuit would succeed depends, however, on the applicable law and its substantive rules.

The Brussels I Regulation and the Lugano Convention 2007 provide for a number of exceptions from the jurisdiction at the defendant's seat. The exceptions that could be relevant in the present context depend on the qualification of the respective activity as contractual or tortious.[75] Claims against activities that add to global warming and climate change and thereby threaten to cause damage to others will in most cases probably sound in tort and are thus to be classified as tortious. On the other hand, contractual claims also can be imagined, for instance if an insurance company requests its insured – in accordance with the conditions of the insurance contract – to refrain from specific climate-sensitive activities the insured is undertaking or – probably more realistically – the claim of the insured against the insurance company's refusal to cover under such a contract damage arising from such climate-sensitive activities.

For claims in tort Article 5(3), Brussels I Regulation/Lugano Convention 2007 is the relevant jurisdictional provision. It is settled that also claims for injunctive relief against threatened damage fall under this provision.[76] This was the purpose of adding the phrase "or may occur" to the former formulation "where the harmful event occurred." Therefore

69 Art. 2(1) Brussels I Regulation; Art. 2(1) Lugano Convention 2007.

70 Art. 59(1) and (2) Brussels I Regulation/Lugano Convention 2007.

71 Art. 60(1)(a)-(c) Brussels I Regulation/Lugano Convention 2007.

72 Art. 5(5) Brussels I Regulation/Lugano Convention 2007.

73 *See* European Court of Justice [2000] ECR I-5925 (C-412/98, *Group Josi Reinsurance v. Universal General Insurance*) for a claim of a Canadian against an EU-seated insurance company.

74 D. Acocella, *in* A.K. Schnyder (Ed.), *Lugano-Übereinkommen zum internationalen Zivilverfahrensrecht – Kommentar*, Zürich/St. Gallen, Dike Verlag, 2011, Art. 2, para. 9.

75 *See* Art. 5(1) and (3) Brussels I Regulation.

76 *See* ECJ [2002] ECR I-8111 (C-167/00, *Verein für Konsumenteninformation v. Karl Heinz Henkel*); BGH NJW 2006, 689.

the claimant can sue the defendant at the place "where the harmful event occurred or may occur."[77] The second alternative ("may occur") is especially relevant for preventive actions/ injunctions in environmental matters.[78]

The expression in Article 5(3) comprises both the place where the tortfeasor acted or where the immediate damage occurred or threatens to occur.[79] The victim may choose between these places. However, this exception applies only where the defendant has its domicile in an EU (or Lugano) Member State and the forum – place of conduct or damage – is in another EU (or Lugano) Member State.[80] It does not concern the jurisdiction of courts outside the EU/Lugano regime whose jurisdiction rules neither the EU nor the Lugano States are empowered to regulate.

Where the claimant requests an omission Article 5(3), Brussels I Regulation/Lugano Convention 2007 allows suing in any (EU/Lugano) country where the defendant should have omitted its activity. Yet, where the defendant ought to omit a specific activity, jurisdiction is founded in the EU/Lugano Member State where the defendant already performs the activity or intends to perform it or where the damage will most probably occur.[81] Where the claimant requests the reduction or omission of CO_2 emissions, the place is decisive where the defendant did or would emit this greenhouse gas (GHG). For jurisdictional purposes this place must be located within the EU/Lugano area.

In consequence, natural or legal persons domiciled in the EU can be sued at their seat or at the place in the EU where they acted or where the claimant suffered damage or is threatened to suffer damage. The Regulation covers all these cases.[82] For a jurisdiction elsewhere, the rules of the Brussels I Regulation are of no avail. The same applies *mutatis mutandis* under the Lugano Convention 2007.

In sum, claimants from inside and outside the EU/Lugano area can base their applications for injunctive relief on the jurisdiction provisions of the Brussels I Regulation/Lugano Convention 2007, where sufficient contacts to the EU/Lugano area exist.

Provisional Measures

The Brussels I Regulation and the Lugano Convention 2007 contain a specific article on provisional legal measures. Article 31 allows a party to approach any EU/Lugano court for such "provisional, including protective, measures" as foreseen in the law of that court's state

77 On 14 December 2010 the Commission tabled a proposal for an amended Brussels I Regulation (COM [2010] 748 final). However, Art 5(3) remains essentially unaffected.

78 *See* P. Mankowski, *in* U. Magnus & P. Mankowski (Eds.), *Brussels I Regulation* (2nd edn), Munich, Sellier European law publishers, 2012, Art. 5, para. 258.

79 ECJ [1976] ECR 1735 (21/76, *Bier v. Mines de Potasse d'Alsace*).

80 Contrary to the Amendment Proposal for Art. 5 Brussels I Regulation, the new version upholds the requirement of the claimant's domicile within the EU; *see* Art. 7 new Brussels I Regulation.

81 *See* Mankowski, *in* Magnus & Mankowski, 2012, Art. 5, para. 261.

82 *See* European Court of Justice [2000] ECR I-5925 (C-412/98, *Group Josi Reinsurance v. Universal General Insurance*) for a claim of a Canadian against an EU-seated insurance company.

even if the Regulation assigns jurisdiction as to the substance of the matter to the courts of another EU/Lugano Member State. The provision is a further jurisdiction rule and extends jurisdiction for provisional measures: where national law so allows, the claimant can institute proceedings also there – in addition to the courts that are competent under the Regulation[83] or the Convention and even where a lawsuit may already be pending in another court.[84]

In consequence, the claimant can choose to apply for provisional procedural measures either in the court that has jurisdiction under the Regulation for the substance matter or in the court that is competent according to the law of that court's country. It must be noted that the claimant may make use of even those exorbitant national jurisdiction provisions that the Brussels I Regulation and the Lugano Convention explicitly ban[85] such as the jurisdiction under Article 14 and 15 French Code civil (jurisdiction based on French citizenship) or under § 23 German Code of Civil Procedure (jurisdiction based on the mere presence of property of the defendant in the court's country).[86]

Again, the rule is applicable only if the relevant courts are located in the EU or in a Lugano Member State. Moreover, the ECJ requests "the existence of a real connecting link" between the dispute and the forum.[87] Only if national law meets this condition are courts competent under Article 31 Brussels I Regulation/Lugano Convention 2007 to grant provisional relief in accordance with their national law (the national requirements for provisional measures vary considerably among the EU/Lugano Member States).[88] The real connecting link requirement further limits the choice of claimants between the court of the substance matter and other courts for provisional relief.

The term 'provisional measures' is not defined in the Regulation/Convention or the accompanying materials. It comprises those measures that according to the ECJ "are intended to preserve a factual or legal situation so as to safeguard rights the recognition of which is sought elsewhere from the court having jurisdiction as to the substance of

83 *See* European Court of Justice [1998] ECR I-7091 (C-391/95, *Van Uden Maritime BV v. Kommanditgesellschaft in Firma Deco-Line*); further M. Pertegás Sender, *in* Magnus & Mankowski 2012, Art. 31 para. 11; J. Kropholler & J. von Hein, *Europäisches Zivilprozessrecht* (9th edn), Frankfurt/Main, Recht und Wirtschaft 2011, Art. 31 para 10; M. Bogdan, 'The Proposed Recast of Rules on Provisional Measures under the Brussels I Regulation', *in* E. Lein (Ed.), *The Brussels I Review Proposal Uncovered*, London, British Institute of International and Comparative Law, 2012, p. 129; S. Leible, 'Art. 31 para 14', *in* T. Rauscher (Ed.), *Europäisches Zivilprozessrecht – Brüssel I-VO, Brüssel IIa-VO* (2nd edn), Munich, Sellier European law publishers, 2006.

84 European Court of Justice [1998] ECR I-7091 (C-391/95, *Van Uden Maritime BV v. Kommanditgesellschaft in Firma Deco-Line*) paras 29, 34; Kropholler & Von Hein, 2011, Art. 31 para. 14.

85 *See* European Court of Justice [1998] ECR I-7091 (C-391/95, *Van Uden Maritime BV v. Kommanditgesellschaft in Firma Deco-Line*) paras 42; P. Vlas, *in* Magnus & Mankowski 2012, Art. 3, para. 2; Kropholler & Hein, 2011, Art. 31 para. 17.

86 *See* Art. 3(2) in connection with Annex I Brussels I Regulation. The Annex I lists all national jurisdiction rules forbidden under the Regulation.

87 European Court of Justice [1998] ECR I-7091 (C-391/95, *Van Uden Maritime BV v. Kommanditgesellschaft in Firma Deco-Line*).

88 *See* Bogdan, 2012, p. 129.

the matter."[89] The measure must have a provisional character and must not in fact finally decide the dispute or create irreversible consequences.[90] A provisional injunction to stop the emission of CO_2 beyond a certain threshold until a final decision would probably fall under Article 31 Brussels I Regulation/Lugano Convention 2007, whereas the order to enact a certain statutory measure for the protection of the environment most likely would not because it is already the final measure that is sought.

Thus, a claimant either from inside or from outside the EU/Lugano States could make use of Article 31 Brussels I Regulation/Lugano Convention 2007 for interim relief in all EU/ Lugano Member States, depending on the relevant possibilities of the respective national law and on the existence of a real connecting link of the dispute with the forum.[91]

2.6.2.2 Joint Defendants

The Brussels I Regulation and the Lugano Convention 2007 (Article 6(1)]) allow a claimant to sue several defendants in the courts of the country where one of them is domiciled.[92] This can facilitate litigation against several defendants from different countries. However, the respective provision requires that the claims are "so closely connected that it is expedient to hear and determine them together to avoid the risk of irreconcilable judgments resulting from separate proceedings."[93] It further applies only where the relevant domicile of all defendants is located in an EU/Lugano Member State.[94] The provision creates the opportunity to concentrate claims against various defendants in one court if the claims are sufficiently closely related so that their separate decision would risk divergent judgments.

It was disputed whether or not defendants who are domiciled in non-EU States can also be joined in proceedings where only one or more co-defendants are domiciled in the EU. A number of authors supported this possibility;[95] others denied it.[96] Already in *Réunion européenne*,[97] the European Court of Justice refused the application of Article 6(1) in a case where a claimant intended to join a defendant domiciled in one EU State

89 *See* European Court of Justice [1992] ECR I-2149 (C-261/90, *Reichert and Kockler v. Dresdner Bank AG*).

90 European Court of Justice [1998] ECR I-7091 (C-391/95, *Van Uden Maritime BV v. Kommanditgesellschaft in Firma Deco-Line*); European Court of Justice [1999] ECR I-2277 (C-99/96, *H H Mietz v. Intership Yachting Sneek BV*); further Pertegás Sender, *in* Magnus & Mankowski, 2012, Art. 31, para. 22.

91 For instance, under German law provisional relief relevant here is the *einstweilige Verfügung* (provisional order; § 935 Civil Procedure Code).

92 Art. 6(1) Brussels I Regulation/Lugano Convention.

93 Art. 6(1) Brussels I Regulation.

94 *See* Muir Watt, *in* Magnus & Mankowski, 2012, Art. 6, para. 20.

95 *See* for instance Kropholler & Von Hein, 2011, Art. 6, para. 7; Leible, *in* Rauscher, 2006, Art. 6, para. 7; H. Schack, *Internationales Zivilverfahrensrecht* (5th edn), Munich, Beck, 2010, para. 360.

96 *See* Oberlandesgericht Hamburg, *Deutsche Rechtsprechung auf dem Gebiete des Internationalen Privatrechts* 1992 No. 193; H. Gaudemet-Tallon, *Compétence et exécution des jugements en Europe – Règlement n° 44/2001, Conventions de Bruxelles et de Lugano* (3rd edn), Paris, 2002, para. 223; S. Corneloup & R. Hausmann, *in* T. Simons & R. Hausmann (Eds.), *Brüssel I-Verordnung*, Munich, IPR Verlag, 2012, Art. 6, para. 22.

97 European Court of Justice [1998] ECR I-6511 (C-51/97, *Réunion européenne SA v. Spliethoff's Bevrachtingskantoor BV and Master of the vessel "Alblasgracht 002"*).

in proceedings in another EU State against a party that was not domiciled in the EU. The Court reasoned that otherwise the EU-domiciled defendant would lose the protection under the Regulation[98] to defend himself at his place. This decision and also the wording of Article 6(1) that is addressed towards persons "domiciled in a Member State" already militate in favour of a restrictive interpretation of the provision. A further argument in this direction derives from the new Brussels I Regulation. Whereas the Amendment Proposal suggested the deletion of the words "domiciled in a Member State" in Article 6(1), the new provision – now Article 8(1) – maintains exactly the old wording. The proposed amendment would have opened the door to a wider interpretation of the provision, which under policy considerations is clearly preferable because it treats EU- and non-EU domiciliaries on an equal footing. However, the drafters of the revised Regulation evidently refused this idea. Very recently, the ECJ explicitly rejected the wide interpretation of Article 6(1) Brussels I Regulation/Lugano Convention.[99] Only where the co-defendants are domiciled either in other EU or Lugano States Article 6(1) applies and offers the possibility to join all defendants in one court if the ground of claim is sufficiently closely connected.

The close connection Article 6(1) Brussels I Regulation/Lugano Convention 2007 further requires is present where, as the text indicates, a separate treatment of, and decision on, claims creates the danger of irreconcilable judgments. In a case where companies of the same group in a concerted action had violated the claimant's patent in different EU Member States, the ECJ nonetheless denied the close link between the claims because different laws applied to the claims.[100] The decision has been rightly criticized as "overly dogmatic."[101] More recent decisions of the ECJ interpret the close connection element as being much less "dogmatic." The Court does no longer formally require as an indispensable precondition that the legal basis of the different actions or the applicable law must be the same.[102] These are aspects that only the national court has to take into account when deciding whether or not a sufficiently close connection exists. In *Painer v. Standard VerlagsGmbH and others* the European Court of Justice stated:

> Thus, a difference in legal basis between the actions brought against various defendants, does not, in itself, preclude the application of Article 6 (1) of Regulation No 44/2001, provided however that it was foreseeable by the

98 By then it was still the Brussels Convention, which was later replaced by the Brussels I Regulation.

99 ECJ 11 April 2013 (C-645/11, *Land Berlin v. Sapir and others*).

100 European Court of Justice [2006] ECR I-6535 (C-539/03, *Roche Nederland BV v. Frederick Primus and Milton Goldenberg*).

101 Muir Watt, *in* Magnus & Mankowski, 2010, Art. 6 para. 25a.

102 *See* European Court of Justice [2007] ECR I-8319 (C-98/06, *Freeport plc v. Olle Arnoldsson*) para. 40 s; European Court of Justice (C-145/10, *Painer v. Standard VerlagsGmbH and others*) paras 80 *et seq.*, <curia.europa.eu/juris/document/document.jsf?text=&docid=115785&pageInd>.

> defendants that they might be sued in the Member State where at least one of them is domiciled.[103]

It may be inferred from this decision that Article 6(1) Brussels I Regulation/Lugano Convention 2007 allows a concentration of actions in one court where for instance several tortfeasors who are domiciled in different EU/Lugano Member States have caused damage by a concerted act. In that case each tortfeasor must foresee the possibility of being sued in the domicile state of another co-defendant. As mentioned, the ECJ further requests that all tortfeasors are domiciled in the EU/Lugano Member States.

Whether the same rule on close connection should apply in the case of independently acting tortfeasors is, however, doubtful. Taking the ECJ's requirement of foreseeability seriously, this militates against the application of Article 6(1) Brussels I Regulation/Lugano Convention 2007. An independent tortfeasor can normally not foresee that and where other tortfeasors will act and might be sued. The European Court of Justice gives a clear hint in this direction when it states:

> For that purpose <*sc. to avoid the risk of irreconcilable judgments*>, the fact that defendants against whom a copyright holder alleges substantially identical infringements of his copyright *did or did not act independently* may be relevant.[104]

Even if one would not follow the ECJ decision that Article 6(1) Brussels I Regulation/Lugano Convention 2007 requires the domicile of all tortfeasors within the EU/Lugano area, a joint suit would hardly be foreseeable for independently acting tortfeasors domiciled outside this area. For the outsiders it is even less foreseeable that they might be sued in the EU/Lugano area than for EU/Lugano domiciliaries. The EU/Lugano scheme would then develop Article 6(1) Brussels I Regulation/Lugano Convention 2007 into a kind of US-American "long arm statute"[105] if in such a case all tortfeasors could be sued in the EU or the Lugano States because only one of them is domiciled there although all others have no connection whatsoever with that jurisdiction. This would mean an overstretching of the jurisdictional competence of the EU and Lugano States. In any event, the lack of forseeability would prevent the application of Article 6(1) in those cases.

In proceedings concerning injunctions against climate change, it is rather obvious that the potential defendants – all those who allegedly have contributed to the climate change – have

103 European Court of Justice (C-145/10, *Painer v. Standard VerlagsGmbH and others*), para. 81, <curia.europa.eu/juris/document/document.jsf?text=&docid=115785&pageInd>.

104 European Court of Justice (C-145/10, *Painer v. Standard VerlagsGmbH and others*), para. 83 (emphasis added), <curia.europa.eu/juris/document/document.jsf?text=&docid=115785&pageInd>.

105 US long-arm statutes require for jurisdiction only a minimal connection to the forum, for instance the conclusion of a contract in the country of the forum.

not acted in concert but independently. It is unlikely that in these cases the ECJ would accept a special interpretation of Article 6(1) Brussels I Regulation granting a court of an EU Member State jurisdiction on actions against all these defendants and thus accord EU courts in fact worldwide jurisdiction over climate change injunctions. Despite the desirability of a central court for such global questions, procedural justice requires to refuse such a wide-ranging jurisdiction of an arbitrarily chosen national court of first instance. This court wherever situated is not better equipped than any other court and is certainly in no better position to decide the case than the courts at each defendant's domicile. In sum, in the cases here under discussion, Article 6(1) Brussels I Regulation/Lugano Convention 2007 is of no avail unless the defendants acted in concert.

2.6.2.3 Enforcement

Judgments rendered in the EU or in a Lugano Member State can rather easily be enforced in other EU/Lugano States provided the judgment is final.[106] If the judgment has been declared enforceable in the EU/Lugano State where it has been rendered, then an immediate declaration of enforcement can be sought in the courts of any other EU/Lugano State where it shall be enforced.[107] Besides, there are only few formalities that must be observed, for instance the production of a copy of the respective judgment.[108] The revision of the Brussels I Regulation has even further facilitated the enforcement procedure by abolishing the requirement of a declaration of enforcement in the country of enforcement (abolition of exequatur).[109]

In general, the same requirements for enforcement apply to provisional injunctions; however, the European Court of Justice here raised additional hurdles. The Court requests that those injunctions must be rendered in proceedings where the defendant party was given the opportunity to be heard.[110] Otherwise these injunctions cannot be recognized and enforced in other Member States. Further, where the injunction already orders a performance, it must be secured that the eventual return is guaranteed.[111] Provisional injunctions that are not rendered under the jurisdiction provisions of the Brussels I Regulation/Lugano Convention 2007 but under exorbitant national jurisdiction rules can generally not be recognized and enforced because they lack a sufficient real connection to the forum.[112]

Also under the Brussels I Regulation/Lugano Convention it must be kept in mind that judgments against other states concerning sovereign acts cannot be enforced.

---

106 *See* Arts. 38 *et seq.* Brussels I Regulation/Lugano Convention 2007.

107 *See* Art. 41 Brussels I Regulation.

108 Arts. 53-55 Brussels I Regulation.

109 *See* Recital 26 and Arts. 39-44 revised Brussels I Regulation.

110 *See* ECJ (125/79).

111 ECJ [1998] ECR I-7122 (C-391/95).

112 Saenger & Dörner, *ZPO* (4th edn) 2011, Art. 31 Brüssel I-VO, para. 18; Kropholler & von Hein, *Europäisches Zivilprozessrecht*, 9th edn, 2011, Art. 31, para. 24.

2.7 Jurisdiction and Enforcement under National Law

Owing to the gaps that the existing international and regional regulations of jurisdiction and enforcement of judgments leave not infrequently, the national rules on these issues will apply.

2.7.1 *General Jurisdiction Rules*

As already mentioned, it is a globally accepted basic principle that the claimant can generally sue the defendant at the latter's seat.[113] Thus, in principle, potential claimants would have to sue each potential defendant at its seat. However, besides that basic jurisdiction rule, national law often grants its courts jurisdiction for injunctive relief under further conditions that vary considerably. The best evidence of these differences is Annex I of the Brussels I Regulation/Lugano Convention 2007. The Annex lists the so-called exorbitant (national) rules on jurisdiction, which the Regulation and the Convention[114] exclude altogether but which can be invoked under national law; they differ from country to country.[115] Moreover, the national rules may provide for further jurisdiction grounds. For instance, national law often allows joining claims of several defendants if there is a close connection between the claims and a procedural need to decide on all claims in one proceeding.[116]

For a suit against potential contributors to the climate change, also the courts of the country where damage threatens to occur will probably have jurisdiction. However, it is a separate and problematical question whether an injunction rendered by such a court would be recognized in the country where the potential contributor is seated. This question will be discussed below.

Whether there is any national law that would allow the concentration of all climate change injunctions in one of its courts appears at best unlikely. It is even doubtful whether the possibility of a class action in the US could cover this case.

Also national law has to respect the immunity of other states where they have acted in their sovereign state capacity (*acta iure imperii*). In such a situation no claim is admissible.

2.7.2 *Specialized Internal Jurisdiction for Environmental Matters*

There is already a number of countries that internally assign environmental issues to specialized environmental ('green') courts that would also be competent for disputes

113 *See* M. Illmer, 'Jurisdiction (PIL)', *in* J. Basedow, K.J. Hopt & R. Zimmermann (Eds.), *The Max Planck Encyclopedia of European Private Law*, Vol. II, Oxford, Oxford University Press, 2012, p. 1021; Schack, 2010, para. 192.

114 Except under Art. 31 Brussels I Regulation; *see* above under 2.6.2.3.

115 The list includes, for instance, the already above-mentioned Arts. 14 and 15 French Code civil or § 23 German Civil Procedure Code.

116 *See*, for instance, § 93(1) Austrian Jurisdiktionsnorm.

concerning the climate change and its effects. A prominent example is India, which enacted a National Green Tribunal Act of 2010. The Act established a special tribunal with jurisdiction "over all civil cases where a substantial question relating to environment (including enforcement of any legal right relating to environment) is involved."[117] Also, for instance, Denmark, Ireland and Japan[118] have established specific court-like bodies for environmental issues.[119]

2.7.3 Enforcement of Foreign Judgments

National law generally erects certain hurdles for the recognition and enforcement of foreign judgments. Some countries require an international treaty with the country where the judgment was rendered;[120] others request reciprocity,[121] yet others require certain minimum conditions.[122] The picture is far from being uniform. Often the recognition and enforcement of a foreign judgment is doubtful or even refused.

As a matter of principle, judgments against other states concerning acts of sovereign power cannot be enforced. This means that judgments against states can only be rendered and enforced in a state by its own courts according to the respective national law. Moreover, even if such a possibility exists, whether a court is prepared to rule against the own state is not always certain.

On the internal national level, final or provisional injunctions of a court can be enforced in that country by the normal means of execution that apply there. Even an injunction granted against the own state could be executed. Partly the means of execution may allow performance of the required act in kind;[123] partly they do not, but grant instead a money sanction or even imprisonment where the debtor does not comply with the injunction.[124]

2.7.4 Neutrality of National Courts

It is a special problem of national courts that their independence and neutrality is not fully safeguarded in every country. Political pressure, bribery and corruption sometimes

117 Sec. 14(1) National Green Tribunal Act 2010.

118 In Japan it is a Commission, the Environmental Dispute Coordination Commission, with quasi-judicial powers that has been established already in 1972; *see* Y. Takamura, 'Japan', *in* R. Lord *et al.* (Eds.), *Climate Change Liability. Transnational Law and Practice*, Cambridge, Cambridge University Press, 2012, p. 233.

119 *See* Pring & Pring, 2009, p. 10 s. The Appendix I of that work contains a list of countries with special environmental courts, tribunals or quasi-judicial bodies (commissions etc.).

120 In general Russia.

121 For instance Germany (§ 328 [1] No. 5 ZPO; except for non-financial cases).

122 For instance Switzerland (Arts. 25-28 Swiss IPRG).

123 *See*, for instance, § 888 German Code of Civil Procedure.

124 *See* § 890 German Code of Civil Procedure.

influence the outcome of court proceedings. Sometimes even a certain bias to serve the national interest may influence a court's decision. All this is generally much less the case, if at all, with international courts.

2.8 Determination of the Applicable Law

If a court is competent to decide an international environmental dispute, the question of the applicable law must be answered. The question concerns which law the court applies to procedural aspects as well as which law it applies to the material law aspects. For the procedural aspects it is generally the forum's own law that is applicable. For the material law aspects, first the rules of private international law must determine the applicable law. Again the distinction between international, regional and national courts is necessary.

2.8.1 International Court of Justice

Both the procedural rules and the substantive law that the ICJ is obliged to apply are provided for by the ICJ's Statute. Articles 39-64 of the Statute and in addition the Rules of Court regulate the procedure; the famous Article 38 of the Statute determines the applicable substantive law.

Article 38 ICJ Statute lists the sources of law that the ICJ may apply and puts these sources at the same time in a specific order: 1) international conventions that are binding for the states involved, 2) international customary law, 3) the general principles of law recognized by civilized nations, 4) judicial decisions and teachings of renowned specialists of international law. From these sources, the Court must distil the legal rules applicable in the case at hand. In the first instance, the applicable international conventions and treaties form the body of law that is to be applied and often to be interpreted for this purpose. But the Court also develops the applicable law. For instance, the Court inferred the principle

> that customary international law provides for restitution as one form of reparation for injury, restitution being the re-establishment of the situation which existed before occurrence of the wrongful act. The Court further recalls that, where restitution is materially impossible or involves a burden out of all proportion to the benefit deriving from it, reparation takes the form of compensation or satisfaction, or even both.[125]

In the view of the Court, restitution in kind, if possible, is evidently the primary remedy where a state wrongfully breaches an international environmental obligation deriving

125 ICJ, *Argentina v. Uruguay* (*Pulp Mills*), para. 273.

from a treaty or from international customary law. The Court has considerable discretion in developing these principles. However, as already mentioned thus far, the Court was reluctant to take the lead in the field of international environmental law.[126]

2.8.2 Applicable Law on the Regional Level

The European regional integration again provides the leading example how the applicable law is to be determined in a regional integration movement.

Procedural aspects that the European Court of Justice has to observe are regulated in Articles 251-281 TFEU and in the Statute of the Court. The ECJ thus follows own procedural rules. The Court may, for instance, order the halt of an allegedly wrongful act or conduct[127] or award – even drastic – money penalties against Member States for non-compliance with EU law.[128] Preliminary injunctive relief is available if waiting for the final decision would cause serious and irreversible damage to the claimant whose interest in preliminary protection must prevail over the interests of the defendant or other involved persons.[129]

On the level of substantive law the ECJ generally applies (and controls the correct application of) specific EU law, which constitutes a separate body of law. In the field of environmental law there are a remarkable number of EU provisions and instruments that directly, or indirectly via Directives,[130] regulate environmental issues and set standards, for instance for the quality of water.[131]

A certain exception to the self-regulation of EU environmental law is the compensation of damage for which the Union is liable. This liability can become relevant in environmental disputes, in particular where the EU omitted to act. Article 340(2) TFEU provides that

> the Union shall, in accordance with the general principles common to the laws of the Member States, make good any damage caused by its institutions or by its servants in the performance of their duties.

This provision somewhat resembles Article 38(3) ICJ Statute ("the general principles of law recognised by civilised nations"). The European Court of Justice had to discover,

126 In *Pulp Mills*, the Court exclusively examined whether Uruguay had violated substantive environmental obligations under the respective treaties with Argentina. The Court could and should have answered as well whether Uruguay had violated environmental duties under international customary law. The Court was evidently of the opinion that it had only jurisdiction as defined by these treaties that contained a submission clause to the ICJ (*see Pulp Mills*, paras 44-68). However, it is more than doubtful whether a submission under a treaty restricts the legal standard of examination and excludes the application of international customary law for the matters covered by the treaty.

127 *See* Art. 278 TFEU.

128 Art. 260(2) TFEU.

129 *See, e.g.*, ECJ [1990] ECR I-3351 (C-195/90; *Commission v. Germany*).

130 The EU Member States are obliged to implement the Directives into their national law.

131 For a good survey *see* Roland *et al.*, 2013, pp. 598 *et seq.*

assemble – and to a certain extent develop – the general compensation principles of the Member States, which are, however, less vague than the principles recognized by civilized nations. The ECJ inferred four requirements from Member States' law: official activity, wrongful act or omission, damage, causal link between official wrongful activity and damage.[132] Like the ICJ, the ECJ need not apply rules of private international law to determine the applicable law.

If the EU Member States do not fulfil their duty to implement and apply EU law, the so-called *Francovich* doctrine of the ECJ steps in. In the case *Francovich a.o. v. Italy*,[133] the ECJ invented the liability of the Member States for non-compliance with EU law. According to this doctrine, an EU citizen can claim damages against his or her State if the EU law grants him or her a subjective right, if the Member State did not observe the EU law, if the citizen suffered damage, if the neglect of EU law was sufficiently qualified[134] and if the damage was caused by the disregard of EU law.[135] Even legislative omissions of implementing EU law bring the *Francovich* doctrine into play.[136] The only sanction is damages; the citizen cannot request restoration in kind. Although EU law has adopted the precautionary principle in environmental matters that aims at prevention,[137] the disregard of environmental EU law by Member States leads to possible compensation if damage occurred.

2.8.3 *Applicable Law on the National Level*

For procedural questions it is a widespread rule that the procedure in principle is ruled by the forum's own law.[138] Although there are exceptions,[139] they do not interest us here. The substantive questions require first the application of the rules of private international law. For international environmental torts, the application of the law of the place where the tort occurred and/or where the damage was suffered is a widely accepted norm.[140] Most of the environmental disputes will arise from non-contractual conduct. Therefore, the law of the country where the tortfeasor acted or the law of the country where the victim suffered the violation of its rights will apply. This is the material

132 *See*, for instance, ECJ [1976] ECR 711 (Case 56-60/74, *Kampffmeyer*) paras 13 *et seq.*

133 ECJ [1991] ECR I-5357 (C-6/90 and 9/90, *Francovich a.o. v. Italy*).

134 The non-implementation of an EU Directive is always a sufficiently qualified disregard of EU law.

135 For a recent account of the ECJ decisions on the Francovich doctrine *see* U. Magnus, *in* L. Tichý & J. Hrádek *et al.* (Eds.), *Odpovědnost Státu za Legislativní Újmu/Staatshaftung für legislatives Unrecht*, 2012, pp. 105 *et seq.*

136 *See* the *Francovich* case itself, where Italy did not implement in time a EU Directive.

137 *See* Art. 191(2) TFEU.

138 *See* Schack, 2010, paras 40 *et seq.*

139 *Ibid.*

140 *See*, for instance, Art. 7 Rome II Regulation, which allows the claimant to choose between the law of the place where the damage was suffered and where the tort was committed.

tort law of the respective country. The substantive tort law differs considerably from country to country in regard to the conditions of liability as well as in regard to the consequences.[141]

2.9 Requirements for Provisional and Final Preventive Measures

2.9.1 *The International Court of Justice – The Global Level*

The ICJ infers from the procedural rules of its Statute and of the Rules of Court that it is entitled to order specific measures as the case may require.[142] The measures may be either **provisional for the period of the proceedings until judgment or final in the judgment.** The provisions of the Statute and the Rules of Court do not prescribe which measures the Court must choose. Thus, the ICJ itself has to develop the principles and boundaries that such measures must comply with.[143] It is an autonomous set of rules given by the Court itself. So far, the Court has established certain requirements. In certain respects, the requirements are still vague or open.

2.9.1.1 Provisional Measures

For provisional measures, Article 41(1) Statute provides:

> The Court shall have the power to indicate, if it considers that circumstances so require, any provisional measures which ought to be taken to preserve the respective rights of either party ("indicate" means to order).

Articles 73 and 75 Rules of Court add that the parties may apply for such measures or that the Court may *proprio motu* indicate, *i.e.* order, such measures. According to Article 76 Rules of Court, the ICJ

141 For the various differences *see* the comparative publications of the European Group of Tort Law, all published by The Hague, Kluwer: J. Spier (Ed.), *Limits of Liability: Keeping the Floodgates Shut*, 1996; J. Spier (Ed.), *The Limits of Expanding Liability. Eight Fundamental Caases in a Comparative Perspective*, 1998; H. Koziol (Ed.), *Unification of Tort Law: Wrongfulness*, 1998; J. Spier (Ed.), *Unification of Tort Law: Causation*, 2000; U. Magnus (Ed.), *Unification of Tort Law: Damages*, 2001; B.A. Koch & H. Koziol (Eds.), *Unification of Tort Law: Strict Liability*, 2002; J. Spier (Ed.), *Unification of Tort Law: Liability for Damage Caused by Others*, 2003; U. Magnus & M. Martín-Casals (Eds.), *Unification of Tort Law: Contributory Negligence*, 2004; W.V. Horton Rogers (Ed.), *Unification of Tort Law: Multiple Tortfeasors*, 2004; P. Widmer (Ed.), *Unification of Tort Law: Fault*, 2005.

142 *See*, for instance, ICJ, *Argentina v. Uruguay* (*Pulp Mills*), Judgment of 20 April 2010, paras 277 *et seq.*

143 *See also*, with respect to provisional measures B. Kempen & Zan He, 'The Practice of the International Court of Justice on Provisional Measures: The Recent Development', *ZaöRV* 69, 2009, pp. 920; *also* S. Rosenne, *Provisional Measures in International Law: The International Court of Justice and the International Tribunal of the Law of the Sea*, 2005, p. 33.

> may, at any time before the final judgment in the case, revoke or modify any decision concerning provisional measures if, in its opinion, some change in the situation justifies such revocation or modification.

The Court has thus wide discretion concerning provisional measures. The measures are binding on the parties;[144] however, the Court can change the ordered measures if the situation has changed. In general, the Court is reluctant to order provisional measures. They are the exception.[145]

With respect to the general requirements the provisional measures must meet, first, it is necessary that the Court's jurisdiction is *prima facie* established.[146] This concerns both the jurisdiction *ratione personae* and *ratione materiae*.[147]

Secondly, a provisional measure requires that there is sufficient probability that the defendant state has infringed an international obligation.

Thirdly, the violation of international law must cause or threaten to cause damage to the right of a party. As the *Pulp Mills* case of the ICJ indicates, it can be sufficient that the damage threatens the environment.[148] However, the issue of causation will often be problematic in environmental matters. This is particularly the case where the core issue is the question whether and to what extent a specific state caused or contributed to or omitted to fight against the global warming and the consequential rise of the sea level. To overcome this hurdle is difficult, though not impossible (as argued elsewhere in this book). For provisional measures the certainty standard of causation may, moreover, be lower than for final measures. Thus far, it is still open how the ICJ decides this question.

Fourthly, the measure must be necessary ("ought to be taken") to preserve the right of a party. Beyond the verbal meaning of this formulation the Court requires that the measure is necessary in order to avoid that during the proceedings a situation of *fait accompli* for the dispute occurs.[149] The need for the measure includes certain urgency which, as the Court puts it, "presupposes that irreparable damage shall not be caused to rights which are the subject of dispute in judicial proceedings."[150] The urgency is also indirectly indicated by Article 74 Rules of Court, which provides that "[a] request for the indication of

144 *See* ICJ *Germany v. United States of America* (*LaGrand*), Judgment of 27 June 2001, paras 102 *et seq.*, 109.

145 In the period from 1947 to 2009 less than half of the applications for provisional measures were successful; *see* Kempen & Zan He, 2009, p. 919, note 1.

146 *See* ICJ Order of 8 April 1993, ICJ Rep. 1993, pp. 3 *et seq.*, para. 14, *Bosnia and Herzegovina v. Serbia and Montenegro* (*Application of the Convention on the Prevention and Punishment of the Crime of Genocide*).

147 *See* ICJ Order of 8 April 1993, *Bosnia and Herzegovina v. Serbia and Montenegro* (*Application of the Convention on the Prevention and Punishment of the Crime of Genocide*).

148 ICJ, *Argentina v. Uruguay* (*Pulp Mills*), Order of 13 July 2006, para. 73.

149 *See* ICJ, *Argentina v. Uruguay* (*Pulp Mills*), Order of 13 July 2006, para. 78.

150 ICJ Order of 8 April 1993, ICJ Rep. 1993, pp. 3 *et seq.*, para. 34, *Bosnia and Herzegovina v. Serbia and Montenegro* (*Application of the Convention on the Prevention and Punishment of the Crime of Genocide*).

provisional measures shall have priority over all other cases." This priority is only justified if there is the imminent danger of important damage. The Court can thus, for instance, order that a state ceases a certain activity that otherwise would cause severe damage and possibly contravene international law.[151] Where the state already stopped its activity, there is no need for a formal order of cessation.[152] In the *Pulp Mills* case the Court could have ordered that Uruguay stopped the construction of the disputed mills until the decision of the case.[153] However, there must be a real threat of damage that could not fully be compensated through the final judgment. In environmental matters a depreciation of the environment will rarely be compensable by later money payments.

The Court has further held that only those measures can be applied for and ordered "which might ultimately form the basis of a judgment."[154] This means that measures unrelated to the dispute cannot be requested; it goes almost without saying that only such measures are admissible that protect rights in dispute between the parties.[155] On the other hand, the final judgment can at least order all those measures that are provisionally admissible.

The provisional measure must also be such that it does not already finally conclude the case. For instance, the dismantling of a mill whose environmentally detrimental operation is the core of the dispute cannot be claimed or ordered on a provisional basis.[156]

Taken together, before the ICJ a successful application for a provisional measure must overcome rather high hurdles.

2.9.1.2 Final Judgments

Neither the ICJ Statute nor the Rules of Court state in detail which measures final judgments can order. It is again the Court's discretion that determines the measures. It appears that the ultimate judgment may always give force to those measures that have been – correctly – provisionally ordered. But evidently, the judgment can go further. It can also order measures that are not admissible as provisional measure because they would already conclude the case. Such – final – measure could be, for instance, the ordering of

151 *See* ICJ Order of 8 April 1993, ICJ Rep. 1993, pp. 3 *et seq.*, para. 52, *Bosnia and Herzegovina v. Serbia and Montenegro* (*Application of the Convention on the Prevention and Punishment of the Crime of Genocide*). The ICJ ordered that "[t]he Government of the Federal Republic of Yugoslavia (Serbia and Montenegro) should immediately [. . .] take all measures in its power to prevent commission of the crime of genocide."

152 ICJ, *Argentina v. Uruguay* (*Pulp Mills*), Judgment of 20 April 2010, para. 269.

153 *See* ICJ, *Argentina v. Uruguay* (*Pulp Mills*), Order of 13 July 2006, paras 73 *et seq.* (no "imminent threat of irreparable damage" by mere construction of the mills whose operation was scheduled to start one year and two years later. The Court stressed, however, that Uruguay had to bear all risks, for instance, of a later dismantling of the mills [para. 78]).

154 ICJ Order of 8 April 1993, ICJ Rep. 1993, pp. 3 *et seq.*, para. 35, *Bosnia and Herzegovina v. Serbia and Montenegro* (*Application of the Convention on the Prevention and Punishment of the Crime of Genocide*).

155 *See* thereto with references Kempen & Zan He, 2009, pp. 919, 923 *et seq.*

156 *See* the facts and claims in ICJ, *Argentina v. Uruguay* (*Pulp Mills*).

the dismantling of a mill, as the ICJ considered in *Pulp Mills*[157] – and refused it as inappropriate remedy for the breach of mere procedural obligations.[158] Thus, the final measure must not be a sanction out of proportion in regard to the violation of international law.
The general requirements for ordering measures in the final judgment are in principle the same as those for provisional measures, but in some respects they are stricter: *Prima facie* jurisdiction no longer suffices; jurisdiction must be firmly established. The same is true for the violation of international law. For the judgment the violation must be fully established. Thus far, the case law of the ICJ does not indicate that a different standard for provisional measures and final judgments applies with respect to the probability with which the damage must threaten. Whether the proof of the necessary causation is stricter for the final judgment than for the provisional order appears to be open.
In the environmental context the restoration in kind is of specific interest; it is probably the – second best – final measure where prevention is no longer possible because the damage already occurred. The ICJ has held that restoration in kind – "re-establishment of the situation which existed before the occurrence of the wrongful act"[159] – is an admissible form of reparation that the Court can order in a final judgment if restoration is possible and not out of proportion to the benefit deriving from it.[160]

2.9.2 *The European Court of Justice – The Regional Level*

The primary task of the European Court of Justice is to safeguard the correct interpretation and application of EU law.[161] This task necessarily includes the environmental enactments of the EU as well as those international conventions that the EU as such has ratified and taken over into EU law, as for instance the Aarhus Convention of 1998.[162] Yet the competence of the ECJ[163] is limited to specific kinds of proceedings that include disputes between EU organs, proceedings with Member States, certain claims of individuals and preliminary rulings on reference of courts of the Member States. It is only within these proceedings that the ECJ can order provisional measures and give judgment. Moreover, the Court can only react to what the parties applied for. In general as well as in particular,

157 In *Pulp Mills* private enterprises had built the mills and Uruguay had granted construction licences for them. The dismantling of the mills would have required administrative action of Uruguay against the enterprises, which then in turn would have been obliged to dismantle the mills.
158 ICJ, *Argentina v. Uruguay* (*Pulp Mills*), Judgment of 20 April 2010, para. 275.
159 *Ibid.*, para. 274.
160 *Ibid.*
161 *See* Art. 19(1) TEU.
162 *See* to its interpretation ECJ 16 February 2012 (C-182/10, *Solvay v. Région wallone*).
163 Besides the European Court of Justice (ECJ), there is also another European Court, the General Court (the former Court of First Instance) with own competences and Rules of Procedure. Under limited conditions an appeal lies against decisions of the General Court to the ECJ (Art. 258 TFEU). The following text focuses on the ECJ.

in the field of environmental law the EU organs and the Member States can sue other organs or Member States for violation of EU law[164] or even for passivity.[165]
For private legal or natural persons the possibility to approach the ECJ is limited.[166] They can attack an act of an EU organ before the ECJ only if the act directly and individually affects them and violates their rights.[167] Because EU law must often be transposed into national law or must be administered by national authorities, it is rather rare that acts of an EU organ directly and individually affect private natural or legal persons. Under restrictive conditions, individuals can also institute proceedings in the ECJ for mere passivity of an EU organ or an EU authority.[168] However, it has to be noted that in a reference procedure[169] the ECJ held that EU law entitles non-governmental environmental organizations to sue even if the underlying EU law protects only the interests of the general public.[170]
Most environmentally relevant decisions of the ECJ concern suits between the EU Commission and a Member State and deal with the interpretation of environmental EU Directives.[171] The core issue is regularly whether the Member State complied with its obligations under the respective Directive.[172] Thus, there are instruments to enforce EU law concerning the environment. However, what EU law does not regulate cannot be enforced.

2.9.2.1 Provisional Measures

Several provisions of the Treaty on the Functioning of the European Union enable the ECJ to grant provisional measures. Articles 278 and 299 TFEU allow the suspension of otherwise executable decisions. The most general provision is Article 279 TFEU, which entitles the ECJ to "prescribe any necessary interim measures" during the proceedings. However, the ECJ can order all interim measures only if a party applied for such a measure.[173] In contrast to the ICJ, the ECJ cannot act *proprio motu*. Thus far, there have been only very few applications for interim measures, and only very few of those applications were successful.[174]

164 *See* Arts. 258, 259, 263 TFEU.
165 Art. 265 TFEU.
166 *See also* N. Reich, *Understanding EU Law* (2nd edn), Intersentia, 2005, pp. 256 *et seq*.
167 Art. 263 TFEU.
168 *See* Art. 265(3) TFEU; in that case the EU must be obliged to release an act that directly and individually affects the claimant; *see* further Stoll & Rigod, *in* Grabitz, Hilf & Nettesheim, *Das Recht der Europäischen Union* (49th Suppl.), 2012, Art. 265, paras 7 *et seq*.
169 *See* Art. 267 TFEU.
170 ECJ 12 May 2011 (C-115/09, *BUND v. Bezirksregierung Arnsberg, Trianel*), para. 59.
171 *See* the collection of ECJ judgments <curia.eu/Common/recdoc/repertoire_jurisp/bull_cee/data/index_B-21.htm>, which contains abstracts of 430 decisions on environmental matters.
172 *See* for instance ICJ [1990] ECR I-851 (C-339/87, *Commission v. Netherlands*); ECJ [2007] I-5939 (C-507/04, *Commission v. Austria*).
173 *See* Art. 83 Rules of Procedure of the Court of Justice.
174 *See* the figures given by Stoll & Rigod, *in* Grabitz *et al*., 2012, Art. 279, para. 2.

An application for a provisional measure must be admissible and founded. It is admissible if a dispute on a substance matter is pendant for which the ECJ *prima facie* would have jurisdiction; if the claimant is entitled to pursue the substance matter;[175] if the measure applied for is admissible in the specific proceedings and does not go further than what is asked for in the substance matter.[176] The application is founded if the measure is urgent to avoid "serious and irreparable damage"[177] and if the measure is necessary because the substance matter is evidently not without merits and the applicant has adduced proof to establish a *prima facie* case for the interim measure.[178] It is a further implied condition that the threatened damage would be caused by the defendant. The difficult problems of causation in environmental matters reoccur here as well.
The hurdles for provisional measures of the ECJ appear to be at least as high as those before the ICJ.

2.9.2.2 Final Measures

Judgments of the ECJ can at best order those measures that the claimant (EU organ, Member State or exceptionally individuals) justifiably applied for. Thus, EU organs and Member States may successfully sue another organ or Member State in the ECJ for not complying with EU law, including EU environmental law.[179] Where the conditions for provisional measures require mere *prima facie* evidence or probability, for instance for the violation of EU law, the conditions for final judgments are generally stricter and require full proof. Generally speaking, the Court can order those measures in its final judgment that the provisions on the different kinds of proceedings foresee (*e.g.* annulment of an act, declaration of violation of EU law, declaration of omission of an act etc.). Other measures cannot be ordered.

2.9.3 The National Level

It is clear from the foregoing text that often only national courts can be approached for injunctive relief. The conditions under which they grant such relief vary from country to country.[180] It must suffice here that German law is taken as an example for the measures that national courts may order either provisionally or finally.

175 Legal and natural persons can only sue if they can show that their own rights are directly endangered by the acts of the organs of the EU (Art. 263 TFEU).

176 *See* Stoll & Rigod, *in* Grabitz *et al.*, 2012, Art. 279, paras 3 *et seq.* with numerous references to the ECJ case law.

177 *See* already ECJ [1963] ECR 271 (275) (Case 25/62 R1, *Plaumann & Co v. Commission*); also, *e.g.*, ECJ [2004] ECR I-8739, para. 36 (C-7/04, *Akzo and Akros*).

178 *See* Art. 82(2) Rules of Procedure of the Court; further Stoll & Rigod, *in* Grabitz *et al.*, 2012, Art. 279, paras 19 *et seq.*

179 *See* Arts. 258, 259, 263, 265 TFEU.

180 For comparative observations concerning provisional measures, *see* Schack, 2010, paras 426 *et seq.*

2.9.3.1 Provisional Measures

It is a characteristic under German law that the claim that the provisional measure shall safeguard must be plausible and that insofar *prima facie* evidence generally suffices. On the other hand, there must be a need and certain urgency for the provisional measure.[181] This means that without the provisional measure a substantial disadvantage must threaten and the interests of the applicant prevail over those of the defendant.[182]

Provisional injunctive relief against activities that contribute to climate change could only be ordered if the claimant can plausibly establish a respective claim. In theory, such claim can be based on §§ 1004 and 823 BGB if there is a serious threat of a violation of the rights that are protected under German tort law such as life, bodily integrity, property etc.[183] The threat that for instance the land will be flooded is a sufficiently serious threat to the owner. However, again causation is necessary and problematic. The causal link must connect the contested activity and the threatened violation of rights. The respondent must be responsible for the activity, and the activity must be the cause of the threat.

2.9.3.2 Final Measures

Injunctive relief in a final judgment also requires that a claim can be based on §§ 1004, 823 BGB.[184] Contrary to provisional injunctive relief, *prima facie* evidence does not suffice; full proof in accordance with the general standard is necessary, in particular for the causal link. Since a future violation of a right cannot be proved, objective facts must be proved that indicate the serious threat of the violation.[185]

Thus, under national law, injunctive relief is possible if the specific requirements can be proved.

2.10 CONCLUSIONS

Presently, no international court with full international jurisdiction on environmental matters exists. The ICJ is competent only for claims by states and only against those states that have submitted to the ICJ's jurisdiction; this is a minority of states. The states that are most relevant for the global climate have not submitted to the Court's jurisdiction.

181 *See* §§ 917, 935, 940 German Civil Procedure Code (ZPO) in civil law matters and §§ 80(5), 123 Administrative Courts Ordinance (VwGO) in public law matters.

182 *See* for civil law matters § 935 ZPO (a provisional measure is admissible if "it must be feared that by a change of circumstances the use of a right would be made impossible or much more difficult"); for public law matters *see, e.g.*, BVerwG 13 June 2007, *Neue Zeitschrift für Verwaltungsrecht*, 2007, 1207.

183 *See* BGH NJW 2004, 3701 (3702); Palandt & Bassenge, BGB (72nd edn), 2013, § 1004, para. 32.

184 So-called *vorbeugende Unterlassungsklage* (preventive omission claim).

185 *See* thereto K.H. Gursky, *in* Von Staudinger, *Kommentar zum Bürgerlichen Gesetzbuch mit Einführungsgesetz und Nebengesetzen*, 2006, § 1004, para. 241.

Neither those states nor enterprises can be brought before the ICJ. Even if states have submitted to the ICJ's jurisdiction, there are rather high hurdles for an adjudication that grants injunctive relief against activities by which states contribute to, or better omit to fight against, climate change. On the level of substantive law it is primarily the problem of causation that may hinder any adjudication. Furthermore, the Court's judges have not been selected for their specific experience and knowledge in environmental matters. Finally, the ICJ's judgments cannot easily be enforced, in particular not as far as the direct enforcement of injunctive relief is concerned. For environmental matters the ICJ's jurisdiction suffers from considerable shortcomings. In sum, the court has no direct jurisdiction over all states; it is not at all competent for claims of or against private persons. The ICJ judges have no special expertise in environmental matters. Moreover, the substantive law that the Court applies is primarily based on conventions, international customary law, general principles; recognized practice and doctrine serve as further source. It is a complicated mixture of rules that the ICJ has to take into account. The outcome of its application is rarely clearly foreseeable.

On a regional level the example of the EU demonstrates that the regional unification or at least harmonization of the provisions on jurisdiction and enforcement of judgments helps to improve the possibilities for injunctive relief, although this relief is still regionally restricted and cannot be granted against states if they acted in their sovereign power. The problem of causation remains also here. Moreover, no specific environmental qualifications of the judges of the European Court of Justice are required. The ECJ exclusively interprets and applies regional EU law, which, however, only after five decades begins to form a structured and apprehensible body of law that allows foreseeable solutions.

On the national level, international jurisdiction will often be granted – except against states. The enforcement of national judgments in foreign states remains in many cases doubtful or difficult. On the national level, one eventually finds specialized environmental courts. The independence, impartiality and neutrality of national courts are not secured in every country. The applicable national laws differ widely.

Seen from the global perspective, which the global climate change requires, this present state of affairs is at least unsatisfactory. It does not sufficiently allow bringing those states to court that neglect their environmental duties. Nor can private actors who contribute to climate change and threaten to damage others in other countries always be effectively prohibited from doing that. The legal remedies are thus far insufficient.

2.11 Proposal for an International Environmental Court

To improve the situation, one could imagine a whole bundle of measures on all levels discussed above.

2.11.1 *Improvements on the ICJ Level*

First, one could suggest the removal of the mentioned limitations that thus far restrict the ICJ's relevance in international environmental issues. This could in particular mean to introduce 'compulsory' jurisdiction, namely to extend the Court's jurisdiction also over states that have not voluntarily submitted to the ICJ's jurisdiction. Further, for environmental disputes the desirable special environmental qualification of ICJ judges could be requested and safeguarded. Finally, the means of enforcement of the ICJ's judgments could be improved.

However, there is no need for a general extension of the 'compulsory' jurisdiction of the ICJ – which practically means the abolition of the jurisdictional immunity defence and which would drastically diminish the range of the principle of the states' sovereignty. An urgent need for an extension of the ICJ's jurisdiction exists in the field of international environmental law and maybe in two or three other fields where the activities of one state can and do also affect the whole globe. In environmental matters this is often the case: if for example Brazil would totally destroy its rainforest, this would affect the world climate as a whole.

With respect to the specific qualification in environmental law, again, this would be necessary not generally for ICJ judges, but only if they have to deal with environmental disputes. There are thus specialities that have to be taken into account for environmental disputes and that are not needed for other disputes before the ICJ. The recognition and effective observance of these specialities in a court generally dealing with disputes other than environmental ones is difficult if not impossible. Amendments of the ICJ Statute in the indicated sense would probably not, at least not fully, achieve the desired effect. It is telling that the special Environmental Chamber which the ICJ had established between 1993 and 2006 was never used.

2.11.2 *Improvements on the Regional and National Level*

Secondly, one could think of an improvement of the possibilities regional or national courts dispose of. Here too it could be suggested that the restricted jurisdiction over states be extended, that the expertise and specialization of courts in environmental matters be improved, that a centralization of environmental suits be enabled, that the recognition and enforcement of foreign judgments on environmental issues be improved.

To achieve these improvements, again, would be rather difficult and in particular time consuming. It would need decades to modify, let alone abolish the time-honoured maxim of state immunity for *acta jure imperii* in the field of environmental law. This cannot be achieved on a regional or national level except perhaps in the distant future. For a change within a reasonably short period – which appears urgently needed to avoid a disastrous climate change – a global initiative is necessary. The institution of better environmental expertise and specialization of courts on the regional and national level would help where it occurs. But there is no safeguard that it occurs, and even if so it is more than unlikely that all regions and states will introduce this at the same time and in the same way. The same is true for any centralization of environmental suits and the facilitation of the recognition and enforcement of foreign environmental judgments. It is more than likely that each region and each state would pursue its own concept that in the end the differences between them would continue and impair the effective protection of the environment by legal means, in particular by way of injunctive relief.

2.11.3 The Quest for an International Environmental Court (IEC)

A better and faster remedy for the mentioned shortcomings would be an IEC of its own. Not all but many of the shortcomings of the present regulation could be removed by such a court. For this reason, for almost two decades a growing number of voices have requested the establishment of such a court.[186] Thus far, no effective steps in this direction have been undertaken. Yet it is high time that the global challenge of environmental problems that do not end at national frontiers was no longer dealt with primarily by national courts but by an international court with specific environmental knowledge and expertise. In the following I discuss the different aspects that have to be considered when instituting such a court.

2.11.3.1 Organizational Framework

Because of the global importance of international environmental issues and in particular of the global dimension of the climate change, the proposed IEC needs to be an organ of the United Nations. Therefore, it should be established in addition to the existing ICJ, which is the highest judicial organ and authority of the UN. The IEC should have the same rank and authority as the ICJ, though it need not necessarily be located at the same venue as the ICJ. It should be a specialized court with specialized judges and with a special jurisdiction only for international environmental matters. Since the experiment with a special Environmental Chamber of the ICJ failed for the reasons considered above, the IEC should be organized as a self-standing separate court.

186 *See*, for instance, Postiglione, 1996; Rest, 1998; K.F. McCallion & H. Rahan Sharma, 'Environmental Justice without Borders: The Need for an International Court of the Environment to Protect fundamental Environmental Rights', 32 *Geo. Wash. J. Int. L. & Econ.*, 1999-2000, pp. 351-365; Hey, 2000; Stephens, 2006; D.A. Farber, 'Apportioning Climate Change Costs', 26 *UCLA J. Environmental Law & Policy*, 2008, pp. 21, 34; Stephens, 2009; P. Birnie, A. Boyle & C. Redgwell, *International Law and the Environment* (3rd edn), New York, Oxford University Press, 2009; Akhtarkhavari, 2010; but critical for instance: S.D. Murphy, 'Does the World Need a New International Environmental Court?', 32 *Geo. Wash. J. Int. L. & Econ.*, 1999-2000, pp. 333-349 ; O.W. Pedersen, 'An International Environmental Court and International Legalism', 24 *J Environ. L*, 2012, pp. 547-558 .

The establishment of such a court requires an international convention on the establishment and the statute of that court. This convention had to be negotiated and concluded within the framework of the United Nations. The Statute could be formulated similarly to that of the ICJ.

Even after a convention on the establishment of an IEC would have been concluded, it had to be ratified by the UN Member States, and its entry into force would depend on a sufficient number of ratifications. One could request the ratification by a majority of the 193 UN Member States. However, this could delay the entry into force for a long time. Therefore, a smaller number – between 20 and 50 States – should suffice.

2.11.3.2 Jurisdiction

A crucial issue is the scope of jurisdiction of the IEC. The Court should be competent to decide on environmental matters that have an international dimension. The environmental issue should thus not affect merely one country, but have the potential to also affect other countries. It should not be necessary that another country is already actually affected; however, it must be plausible that the environmental issue, for instance the planning of a highly hazardous plant close to the border, could affect the interests of the neighbouring (and other) countries and their citizens.

The Court would have jurisdiction only for environmental matters which had to be defined for this purpose. For their definition, the Statute of the proposed Court could perhaps partially borrow from the Indian National Green Tribunal Act 2010. This Act grants jurisdiction "*over all civil cases where a substantial question relating to environment [. . .] is involved.*"[187] According to the Act, a "substantial question relating to environment" is not defined exhaustively but shall include two situations: either the violation of environmental law provisions that has relevant consequences or an activity or source with environmental consequences.[188] From this definition the principle could be borrowed that the Court could be seized of instances where existing environmental law has been infringed and serious consequences (concerning health or property) for citizens of more than one country have been caused or are threatening or where environment-sensitive activities have caused or threaten to cause such consequences.

Contrary to the Indian Act, the jurisdiction of the IEC should not confine itself to civil law cases. The jurisdiction should also comprise public law cases since the borderline between civil and public law is not everywhere drawn in the same way. Its complicated distinction

187 Sec. 14(1) National Green Tribunal Act 2010.

188 *See* sec. 2 (m) National Green Tribunal Act 2010: "'A substantial question relating to environment' shall include an instance where, – (i) there is a direct violation of a specific statutory environmental obligation by a person by which, – (A) the community at large other than an individual or group of individuals is affected or likely to be affected by the environmental consequences; or (B) the gravity of damage to the environment or property is substantial; or (C) the damage to public health is broadly measurable; (ii) the environmental consequences to public health relate to a specific activity or a point source of pollution."

should be avoided. However, criminal law cases concerning the environment should not be covered. They should be treated either by national courts or by the International Criminal Court as far as this court is competent.

With regard to the relationship to the jurisdiction of the ICJ, one could imagine that the IEC should have exclusive jurisdiction in the sense that international environmental disputes could not be brought before the ICJ but only before the IEC. It is, however, also imaginable that the claimant should have a choice between the two courts and that the choice once made is then binding on the parties.

2.11.3.3 Selection of Judges

It is essential that the judges of the IEC possess special knowledge and experience in environmental matters. Their high qualification in this field is absolutely necessary. Moreover, knowledge of and experience in international law are also essential. Further, the judges must be qualified under their home countries' requirements to serve at the highest national court or courts.[189]

A related question is the procedure of how and by whom the judges have to be chosen. The judges should be elected by the General Assembly of the United Nations. Unlike with the election of the judges of the ICJ, the participation of the Security Council is dispensable.[190] The Security Council is specialized in political, not environmental, matters. International environmental matters are of a different character than other cases the ICJ is regularly seized with. They often, or in case of climate change problems always, concern the international community as a whole and have much fewer political implications than the 'normal' cases of the ICJ, where rival states ask for a decision of their dispute.

The election must be preceded by a nomination procedure that leads to a list of candidates from which the judges have to be elected. The nomination procedure could follow the rules laid down in Articles 4-7 Statute of the ICJ, which takes account of the different country groups within the UN. One could also think of a procedure that only two or more UN Member States together can nominate one candidate who must be a national of none of the nominating states.

2.11.3.4 Right to Seize the Court

As with the ICJ, each UN Member State should be entitled to seize the IEC. It is suggested here that also a qualified number of natural persons should have the right to institute proceedings. Sometimes states do not react to environmental problems; therefore, a qualified number of persons – for instance one million – should have that right, too. Further,

189 This does not mean that they are already members of one of the highest courts of the country but that they could be selected for this position.

190 The ICJ judges (five judges every third year) are elected by a combined vote of the General Assembly and the Security Council. In each body the candidate must have a majority; *see* Arts. 4(1), 8, 10(1) ICJ Statute.

internationally recognized environmental organizations should be entitled to bring claims before the court. These two further kinds of claimants take account of the fact that serious environmental issues are often raised by private initiatives and that the civil society takes a special interest in the defence of a liveable environment.

One could further imagine that a public prosecutor should be installed with the authority to institute proceedings in the IEC. The draft for an ICE that the private group ICE coalition has prepared proposes such a model.[191] This idea follows the model of the International Criminal Court.[192]

It is here suggested not to follow this example. The institution of a public prosecutor is apt for criminal law courts. It is not suitable for adversary proceedings with which the IEC generally would be seized. This Court should neither interfere with the activities of the International Criminal Court (ICC) nor insofar copy its procedure. For matters like injunctive relief or damages claims, there is evidently no need for a prosecutor. Moreover, this authority that had to have a global scope of competence does not yet exist, and it would be very difficult to establish it on an effective global scale.

2.11.3.5 Possible Defendants

Problematic is also the question of who can be sued. Before the ICJ, only states that have submitted to the Court's jurisdiction can be sued. Under the Rome Statute of the International Criminal Court even situations in states that are not party to the Rome Statute and have thus not submitted to this Court's jurisdiction can be prosecuted if the Security Council acting under Chapter VII of the UN Charter refers the case to the ICC's Prosecutor.[193] The Rome Statute thus allows neglecting the otherwise unimpeachable principle of sovereignty in certain situations. It must be situations where the committed crimes are so serious that the international community can no longer stand by and accept the sovereignty of the involved state as a shield against interference.[194]

For the IEC, it is suggested here that not only states that have ratified the Court's Statute or have submitted to the jurisdiction of the Court can be sued. Where environment-sensitive activities of or from a non-submitting state seriously affect or threaten to seriously affect other countries, the international community must be able to react. Again, the principle of sovereignty should no longer be a shield or excuse in such a situation. The question is, however, who should be allowed to bring such a situation before the Court. As with the ICC, the Security Council could be the organ entitled to this step. However, as already

191 Ch. 2 Art. 1(4) Draft.

192 Art. 15 Rome Statute.

193 Art. 13(b) Rome Statute.

194 The ICC's jurisdiction ratione materiae comprises anyway "the most serious crimes of concern to the international community as a whole" (Art. 5(1) Rome Statute). Thus, the situations referred by the Security Council involve a particularly high level of these most serious crimes.

mentioned, this Council mainly deals with conflicts of a political nature, regularly between two and only very few states. This UN organ is not specialized to deal with environmental questions that concern many if not all UN Member States. It is therefore suggested here that a qualified number of UN Member States should be entitled to institute proceedings against a state that has not submitted to the jurisdiction of the IEC. The qualified number may be either a majority of the General Assembly or, preferably, a lower number of UN Member States – for instance 10 or 30.

It is a further question whether it should be possible to bring private entities also before the IEC. On the one hand it is rather self-evident that not every enterprise could be sued in the IEC. This would overburden the Court in such a way as to make it unable to work. On the other hand, some multinational enterprises have bigger economic power and emit more greenhouse gases than many UN Member States. Their importance for the global climate is evident. They should be possible defendants in the IEC. It is thus necessary to draw a borderline between enterprises that are too small for proceedings in the IEC and those that are big enough. One could make the emission of GH gases the decisive factor. Enterprises that emit 0.1% of all such emissions could be covered.

2.11.3.6 Procedure

For the procedure of the IEC, Articles 39-64 ICJ Statute could be applied accordingly. In particular, as under Article 50 ICJ Statute, the Court must have the right to mandate *suo motu* experts to give expert opinions on environmental issues. The court should not be confined to the evidence presented by the parties.

Like the ICJ, the IEC should be able to establish chambers with a smaller number of judges than are required for the full Court.

2.11.3.7 Applicable Law

The law that the IEC has to apply should be determined in a way that resembles Article 38 ICJ Statute: (1) international conventions that are binding for the states involved should be taken into account, but also other international environmental accords; (2) international customary law with respect to general and environmental matters; (3) the general principles of law recognized by civilised nations; (4) judicial decisions and teachings of renowned specialists of international and international environmental law. Also soft law such as UN declarations etc. on environmental matters should be used, in particular for the examination of a common international conviction and understanding in the interpretation and application of conventions and international customary law.

It is apparent that protection against climate-sensitive activities can only go as far as developed by these sources. Nonetheless, the sources allow the Court certain discretion. The specialized IEC could and should play a more active role in the protection of the environment than the ICJ.

The sources mentioned above secure better than the redress to national law the global perspective of the applicable law. Reference to a specific national law would again nationalize the solutions for international environmental disputes. Even for matters such as injunctive relief or compensation of damage there is sufficient material at least under the general principles of law on which the Court could base its decisions.

2.11.3.8 Possible Judicial Actions

The ICJ Statute does not specify the measures ICJ judgments can order with the exception that Article 41(1) allows the ICJ to "indicate, if it considers that circumstances so require, any provisional measures which ought to be taken to preserve the respective rights of either party." A final ICJ judgment may therefore order what the decision of the respective conflict requires. The same should hold true for judgments of the IEC. This Court should be specifically entitled to grant provisional and final injunctive relief that the defendant must adapt its activities to a certain threshold or refrain from specific activities.

2.11.3.9 Review

The ICJ Statute does not allow for an appeal against an ICJ judgment.[195] The only way of review is a revision if new and decisive facts have been discovered that the party claiming revision did not overlook by negligence.[196]

However, the Security Council of the UN can factually review ICJ judgments when deciding on measures for the enforcement of those judgments and in particular when refusing such measures.[197]

Also judgments of the IEC should be final, binding on the parties and without appeal. However, where a judgment is rendered by a Chamber of the Court an appeal to the full Court should be admitted.

As is the case with ICJ judgments, a revision of a judgment should be possible where new facts are discovered that the claimant was not and could not be aware of and that the Court also did not know.

2.11.3.10 Enforcement of Judgments

A specifically delicate problem is the enforcement of judgments of international courts. A shortcoming of judgments of the ICJ is that, unless they are observed voluntarily, their enforcement depends on the discretionary decision of the Security Council of the UN. As already mentioned, the enforcement of judgments of the ICJ is not regulated by the

195 *See* Art. 60 sent. 1 ICJ Statute: "The judgment is final and without appeal."

196 Art. 61 ICJ Statute.

197 Art. 94(2) UN Charter the Security Council of the UN.

Court's Statute. However, the Security Council can take measures to enforce those judgments.[198] As yet, the Security Council has not often made use of this possibility. Therefore, if not observed voluntarily, the judgments of the ICJ will often not be enforced.

For decisions of the IEC, a more binding mechanism of enforcement is necessary. One could consider that the regulation of enforcement in Articles 103-111 Rome Statute of the ICC could form a model. Under this Statute, the national authorities of a Contracting State must enforce the criminal sanctions levied by the ICC. That means either that they have to enforce fines[199] or if their states have declared their willingness to accept sentenced persons to imprison them.[200] The ICC supervises and controls the enforcement; in case of non-compliance the ICC notifies the respective state or transfers a prisoner to another Contracting State. This procedure is not fully transferable to the enforcement of international environmental judgments. However, the Statute of the IEC should at least provide that the Contracting States are obliged to enforce the judgments of the Court. Yet in case of a state's non-compliance with a judgment, there is no possibility that the Court could transfer the task of enforcement to another state. The non-complying state itself, be it a Contracting State or not, must be forced to enforce the judgment. Direct means of intervention – for instance a direct reduction of GH gases or respective legislation in the non-complying state – are not feasible. The principle of sovereignty prohibits them. Thus, only indirect measures are available that may influence a non-complying state to change its attitude. As with ICJ judgments, one could imagine that the Security Council should be the organ to decide on any measures of enforcement. Here another path is suggested, namely the establishment of a global Environmental Council within the framework of the UN. Neither the Security Council[201] nor the Economic and Social Council[202] nor the Trusteeship Council[203] of the UN has thus far special expertise and qualifications for environmental issues. The Environmental Council of the UN should have only temporary and no permanent members. Moreover, there should be no veto right so that sanctions could be taken if a majority of this Council so decides.

2.11.3.11 Financing

The IEC should charge reasonable fees. Since such fees cannot fully cover the costs of the Court, the additional financing of the Court should be borne by the UN.

198 *See* Art. 94(2) UN Charter.
199 Art. 109 Rome Statute.
200 Art. 103(1) Rome Statute.
201 As to its competence, *see* Art. 24 UN Charter.
202 The competence of this Council is regulated by Arts. 62-66 UN Charter.
203 *See* Arts. 87 *et seq*. UN Charter.

2.11.3.12 The Political Dimension

The project of an IEC has an evident highly political dimension. Without support by a sufficient number of UN Member States such a project would be doomed to failure. Given the urgency that the climate change brings about and the fact that ever more states are affected by it, it appears possible that sufficient support can be gained even in those states that presently might still fear to be successfully sued in the IEC. On the contrary, there is a growing number of states, in particular those mentioned in the introduction to this part, which are highly interested in a forum where they can bring their claims with reasonable prospects of success and enforcement of a respective judgment. They will support the establishing of an IEC.

3 Climate Change and the (Possible) Role of International Criminal Law*

*Gerhard Kemp***

3.1 Introduction and Problem Statement

This contribution is concerned with the possibilities presented by international criminal justice as a response to climate change. It presents a balanced evaluation of international criminal justice mechanisms as possible helpful tools in the broader global response to climate change. I will argue that there are various modalities and mechanisms in the field of international criminal justice that can be useful. However, the pitfalls are also noted.

The inherent characteristics of (international) criminal law as essentially backward looking (that is, reacting to a past event or events) are acknowledged at the outset and can be compared with the arguably more prospective characteristics of other legal responses to climate change addressed in this project, such as the use of injunctions or injunctive relief.[1]

The essential assumption is that climate change is to a large extent caused by human conduct.[2] To the extent that human conduct is *harmful* to the environment, thus causing or contributing to detrimental climate change, the question is whether an (international) criminal justice response or responses would be meaningful, appropriate and effective.

---

* This chapter forms part of a broader project on the interplay between risks posed by climate change and the role of law in response to climate change. Part of the research on this project was conducted at Stellenbosch University (Faculty of Law) and at Stellenbosch Institute for Advanced Study (STIAS) during 2012. The first part of the research was first published in O.C. Ruppel, C. Roschmann & K. Ruppel-Schlichting (Eds.), *International Climate Law and Global Governance: Legal Responses to a Changing Environment*, Nomos Publishers, 2013. The permission to reproduce parts of that chapter for purposes of the current chapter is gratefully acknowledged.

** Professor of Law, Stellenbosch University, on the board of directors, Institute for Justice and Reconciliation, Cape Town, advocate of the High Court of South Africa.

1 *See* Chapter 1.

2 This chapter presents a legal framework and a legal argument; it does not pretend to be a study in natural science. However, respectable voices that support the notion that climate change is to an important degree caused by human activity are referred to, since they seem to represent the majority of scientific opinion on the matter. It is this fundamental observation that forms the basic assumption for purposes of the further legal argument. Compare the following statements on the causes of climate change: J.E. Hansen, Director of the Goddard Institute for Space Studies at NASA, *et al.*, 'Perception of Climate Change', *in The Proceedings of the National Academy of Sciences*, 6 August 2012: "There are many other human-induced stresses on

Philosophically, the question whether an international criminal justice response to climate change is meaningful, or indeed warranted, can perhaps best be answered with reference to Hannah Arendt's distinction between *Verbrechen gegen Menschheit* (crimes against mankind) and *Verbrechen gegen Menschlichkeit* (crimes against humanity).[3] The latter classification is informed by those violations that affect our sensibilities and characteristics as human beings: our sense of private, free, autonomous beings with inherent human dignity. The former group of atrocities affect our very existence and survival. This includes crimes against peace, and should arguably also include crimes against the environment, which, in terms of gravity and scale, constitute threats to the survival of mankind. While crimes against peace (in the form of the crime of aggression) were recognized as the "supreme international crime" at Nuremberg,[4] we have yet to see any comparable criminal justice response to the phenomenon of climate change caused by human conduct.

There are, of course, criminal justice responses to conduct affecting the environment – both under national and international law. For instance, international humanitarian law prohibits widespread, long-term and severe damage to the natural environment. Violations of the relevant rules can lead to individual criminal liability. I will return to the role of international humanitarian law later, but the point here is that the criminal justice response to harmful conduct against the environment is relatively well established. The question is to what extent an (international) criminal justice response to human conduct that is so harmful that it causes climate change, is sensible, and indeed feasible.

---

life [. . .] yet global warming caused by fossil fuel burning may be a unique threat because of the millennial time scale of anthropogenic carbon within surface carbon reservoirs"; R.A. Muller, Professor of Physics at the University of California at Berkeley and Faculty Senior Scientist at the Lawrence Berkeley Laboratory, 'The Conversion of a Climate-Change Skeptic', *New York Times*, 28 July 2012, available at <www.nytimes.com>: "Three years ago I identified problems in previous climate studies that, in my mind, threw doubt on the very existence of global warming. Last year, following an intensive research effort involving a dozen scientists, I concluded that global warming was real and that the prior estimates of the rate of warming were correct. I'm now going a step further: Humans are almost entirely the cause [. . .] How definite is the attribution to humans? The carbon dioxide curve gives a better match than anything else we've tried. Its magnitude is consistent with the calculated greenhouse effect — extra warming from trapped heat radiation [. . .] our analysis does not depend on large, complex global climate models, the huge computer programs that are notorious for their hidden assumptions and adjustable parameters. Our result is based simply on the close agreement between the shape of the observed temperature rise and the known greenhouse gas increase." (Both statements from excerpts of statements on climate change, available at <http://climatechange.procon.org/view.answers.php?questionID=001445> (last accessed on 15 April 2013.)

3 Reference to Hannah Arendt's use of the terms in the reproduction and translation of an interview with the German philosopher K. Jaspers, 'Who Should Have Tried Eichmann?', 4 *Journal of International Criminal Justice*, 2006, p. 855.

4 W.A. Schabas, 'Origins of the Criminalization of Aggression: How Crimes against Peace Became the "Supreme International Crime"', *in* M. Politi & G. Nesi (Eds.), *The International Criminal Court and the Crime of Aggression*, Aldershot, Ashgate, 2004, p. 31.

3.2 International Criminal Justice in Constitutionalist and Global Governance Perspective

The central question in this contribution is whether international criminal justice can contribute meaningfully to efforts to stem global climate change. Margaret Karns and Karen Mingst's neat definition of global governance suggests that international (criminal) law is indeed a "piece of global governance". The "pieces of global governance are the cooperative problem-solving arrangements and activities that states and other actors have put into place to deal with various issues and problems".[5] The most important "pieces of global governance" identified by Karns and Mingst are:

- International law (including international humanitarian law and international criminal law);
- Norms or "soft law" (for instance framework conventions on biodiversity and climate change);
- Formal and informal structures such as intergovernmental organizations, non-governmental organizations, global conferences and so on;
- International regimes (for instance on trade, nuclear non-proliferation, food aid, telecommunication and so on).

If we accept that international law (including international criminal law) is an important part of global governance, the critical question is, under what circumstances should we (the international or global community) resort to international criminal justice as a response to calamities and atrocities?

One view is that the (evolving) system of international criminal justice is primarily a system informed by the international community's *reaction* to atrocities like genocide, crimes against humanity and war crimes. The *ad hoc* nature of international criminal tribunals from Nuremberg to the Yugoslavia Tribunal (ICTY) and the Rwanda Tribunal (ICTR) underscores this view. Of course, the creation of the permanent International Criminal Court (ICC) (with its forward-looking, preventive potential) represents an important turning point away from the primarily reactionary narrative of international criminal justice, to which I will turn to later. The point is that the criminal justice response is by its nature mostly reactionary and backward looking: the emphasis is on punishment (retribution) for past conduct. This is not to say that criminal law theory is one-dimensionally preoccupied with past events: indeed, the utilitarian aspiration of (international) criminal law has an eye on the future as well, namely prevention of further crime, rehabilitation of offenders, and even integration or healing of society affected by crime.[6]

5 M. Karns & K. Mingst, *International Organizations – The Politics and Processes of Global Governance*, Boulder, London, Lynne Rienner Publishers, 2004, p. 4.

6 I. Tallgren, 'The Sensibility and Sense of International Criminal Law', 13 *European Journal of International Law*, 2002, p. 562.

A progressive view is that there are important elements and characteristics of international criminal law that espouse constitutionalist qualities. Constitutionalist in this context means that the international system is moving towards the supranational limitation of state power.[7] In this sense the ICC can also be described as a constitutional development. It impacts on the way states conduct themselves and complements the exercise of state jurisdiction over the most serious crimes under international law.[8]

One should be careful not to view the emergence of international criminal law generally, and the establishment of the ICC in particular, in a too idealistic way. Of course, idealism has always been and will hopefully remain to be an important driving force for good – also with respect to international criminal justice. In practical terms, however, it is prudent to keep in mind that the body of international criminal law (still) consists of two conceptually somewhat different parts, namely international law, with its "diplomatic conferences, convention making by consensus and autopoetic interpretation of law,"[9] and criminal law, which is "supposed to be the, by definition, positivistic discipline of law, based on the fundamental importance of legality, the principle of *nullum crimen sine lege, nulla poena sine lege*."[10] These observations underscore the promises and the pitfalls of international criminal law as a potential tool or 'piece' of global governance, aimed at addressing climate change. In the paragraphs below some of the important promises and pitfalls will be highlighted and analysed in the context of the global governance paradigm.

3.3 The Main Features of the Emerging System of International Criminal Justice

A detailed discussion of the emerging system of international criminal justice falls beyond the scope of this contribution. I will, however, highlight a number of features of what can be called the emerging or evolving system of international criminal justice. In turn, I will indicate the relevance of these features to the topic under discussion, namely international criminal justice as a piece of global governance that might assist with efforts to address global climate change.

---

7 This reflects the development away from the traditionally anarchist international system where sovereign states invariably acted in their own interest. *See* further comments by D. Caron, 'Framing Political Theory of International Courts and Tribunals: Reflections at the Centennial', *ASIL Proceedings*, 2006, p. 56.

8 W. Werner, 'Constitutionalisation, Fragmentation, Politicization, the Constitutionalisation of International Law as a Janus-Faced Phenomenon', 8 *Griffin's View on International and Comparative Law*, 2007, pp. 18-23; G. Kemp, 'Constitutionalization and the International Criminal Court: Whither Individual Criminal Liability for Aggression?', 4 *South African Law Journal*, 2008, pp. 694-714.

9 Tallgren, 2002, p. 562.

10 *Ibid.*, p. 564.

3.3.1 *Individual Criminal Liability*

Individual criminal liability is the cornerstone of modern international criminal law. Indeed, the evolving system of international criminal justice is premised on the notion that the various atrocities that form the subject matter material of international criminal law are committed by persons. In this sense, the Nuremberg and Tokyo precedents[11] established the separateness of international criminal law from other branches of international law. The principle was established that individuals are the subjects of international criminal law and can be held liable for crimes under international law. Decades later this principle was confirmed by the International Criminal Tribunal for the Former Yugoslavia (ICTY) in *Prosecutor v. Tadić*:

> The basic assumption must be that in international law as much as in national systems, the foundation of criminal responsibility is the principle of personal culpability: nobody may be held criminally responsible for acts or transactions in which he has not personally engaged or in some other way participated (*nulla poena sine culpa*).[12]

The Rome Statute of the (permanent) ICC also confirms the principle of individual criminal responsibility as a key part of the Court's jurisdictional regime. Article 25 of the Statute provides as follows:

> 1. The Court shall have jurisdiction over natural persons pursuant to this Statute.
> 2. A person who commits a crime within the jurisdiction of the Court shall be individually responsible and liable for punishment in accordance with this Statute.
> 3. In accordance with this Statute, a person shall be criminally responsible and liable for punishment for a crime within the jurisdiction of the Court if that person:
>   (a) Commits such a crime, whether as an individual, jointly with another or through another person, regardless of whether that other person is criminally responsible;

11 *See* the important observations by Q. Wright, 'The Law of the Nuremberg Trial', *The American Journal of International Law*, 1947, pp. 38-72; T. Taylor, 'The Nuremberg Trials', 55 *Columbia Law Review* 4, 1955, pp. 488-525; H. Leonhardt, 'The Nuremberg Trial: A Legal Analysis', *The Review of Politics*, 1949, pp. 449-476; G. Komarow, 'Individual Responsibility under International Law: The Nuremberg Principles in Domestic Legal Systems', 29 *The International and Comparative Law Quarterly* 1, 1980, pp. 21-37; A. Marston Danner, 'Beyond the Geneva Conventions: Lessons from the Tokyo Tribunal in Prosecuting War and Terrorism', 46 *Virginia Journal of International Law*, 2005, pp. 83-130; C. Carr, 'The Judgment of the International Military Tribunal for the Far East', *Transactions of the Grotius Society, Problems of Public and Private International Law, Transactions of the Year 1948*, 1948, pp. 141-151.

12 *ICTY Prosecutor v. Dusko Tadić (Appeals Chamber)* Case No. IT-94-1-A, 15 Jul [1999], para. 186.

(b) Orders, solicits or induces the commission of such a crime, which in fact occurs or is attempted;
(c) For the purpose of facilitating the commission of such a crime, aids, abets or otherwise assists in its commission or its attempted commission, including providing the means for its commission;
(d) In any other way contributes to the commission or attempted commission of such a crime by a group of persons acting with a common purpose. Such contribution shall be intentional and shall either:
  (i) Be made with the aim of furthering the criminal activity or criminal purpose of the group, where such activity or purpose involves the commission of a crime within the jurisdiction of the Court; or
  (ii) Be made in the knowledge of the intention of the group to commit the crime;
(e) In respect of the crime of genocide, directly and publicly incites others to commit genocide;
(f) Attempts to commit such a crime by taking action that commences its execution by means of a substantial step, but the crime does not occur because of circumstances independent of the person's intentions. However, a person who abandons the effort to commit the crime or otherwise prevents the completion of the crime shall not be liable for punishment under this Statute for the attempt to commit that crime if that person completely and voluntarily gave up the criminal purpose.

No provision in this Statute relating to individual criminal responsibility shall affect the responsibility of "States under international law."

I will return to some of the important implications of the Rome Statute's provisions on individual criminal responsibility when dealing with the ICC as a potential forum to deal with those responsible for climate change.

3.3.2 The Principle of Legality

The notion that legal rules should be clear and certain is not unique to criminal law. But as an element of due process and as a general principle of criminal law, the principle of legality has a central and fundamentally important place in national criminal law as well as international criminal law.[13] There are differences in approach between national and

13 S. Lamb, 'Nullum Crimen, Nulla Poena Sine Lege in International Criminal Law', *in* A. Cassese, P. Gaeta & J.R.W.D. Jones (Eds.), *The Rome Statute of the International Criminal Court, A Commentary*, Vol. I, Oxford, Oxford University Press, 2002, p. 733.

international criminal law, which I briefly discuss below. But first, it is necessary to delineate the normative foundations of the principle of legality in criminal law generally.

In the early 1880s the German legal scholar JP Anselm von Feuerbach coined the maxim *nullum crimen, nulla poena sine lege.* In terms of this principle, no crime (or punishment) can exist without a clear norm in law criminalizing the conduct in question and providing for applicable punishment. The principle thus provides for a crime norm and a punishment norm.[14] In essence, the principle is understood to mean that criminal laws should be made by "a competent legislature that announced in advance and with clarity and certainty the definition of crimes and the details of their punishments."[15]

The principle of legality is firmly established in international criminal law – although with some modifications. It even has customary international law status, which is not surprising, given the widespread and general respect for the principle and international judicial confirmation of its paramount importance.[16] The Rome Statute of the ICC also provides for the principle of legality by splitting it into the two components referred to above, namely *nullum crimen sine lege*, in Article 22,[17] and *nulla poena sine lege*, in Article 23.[18] A further aspect, the non-retroactive application of criminal law, is provided for in Article 24[19] of the Rome Statute.

Although the principle of legality as applied in national criminal law and in international criminal law clearly share the same normative roots (fairness, clear warning that conduct is criminal and so on), it is interesting to note that the principle is sometimes applied less strictly in international criminal jurisprudence. Indeed, the Nuremberg Tribunal set the scene with the retroactive criminalization of and ultimate conviction of the senior Nazis for the crime of aggression, as well as crimes against humanity (war crimes as a distinct category of crimes was already well established). Many commentators objected to the Nuremberg Tribunal's *ex post facto* criminalization of certain conduct.[20] By contrast, and not surprisingly, Telford Taylor[21] argued that international (criminal) law "is not capable of

14 Kemp, 2010, p. 13.

15 J. Burchell, *Principles of Criminal Law* (3rd edn), Cape Town, Juta, 2005, p. 95.

16 G. Werle, *Principles of International Criminal Law*, The Hague, TMC Asser Press, 2005, p. 32.

17 Art. 22: (1). A person shall not be criminally responsible under this Statute unless the conduct in question constitutes, at the time it takes place, a crime within the jurisdiction of the Court. (2). The definition of a crime shall be strictly construed and shall not be extended by analogy. In case of ambiguity, the definition shall be interpreted in favour of the person being investigated, prosecuted or convicted. (3). This article shall not affect the characterization of any conduct as criminal under international law independent of this Statute.

18 Art. 23: A person convicted by the Court may be punished only in accordance with this Statute.

19 Art. 24: (1). No person shall be criminally responsible under this Statute for conduct prior to the entry into force of the Statute. (2). In the event of a change in the law applicable to a given case prior to a final judgment, the law more favourable to the person being investigated, prosecuted or convicted shall apply.

20 *See* the general criticism listed in H. Leonhardt, 'The Nuremberg Trial: A Legal Analysis', 11 *The Review of Politics* 4, 1949, pp. 449-476.

21 Telford Taylor was part of the American prosecution team at Nuremberg. He served on the team of Robert Jackson, who played an important role in the drafting of the Nuremberg Charter.

development by the normal processes of legislation, for there is no continuing international legislative authority." International criminal law "grows, as did the common law, through decisions reached from time to time in adapting settled principles to new situations."[22]
In a famous dissenting opinion delivered at the Tokyo Tribunal, judge Röling put the emphasis not on the pre-existing criminal norm. The judge concluded that crimes against peace were to be punished "because of the dangerous character of the individuals who committed them". The focus was on the danger, rather than on the guilt.[23]
The notion that legality in criminal law can somehow be applied less strictly in international criminal law also found favour at the Special Court for Sierra Leone, which hears cases of crimes against humanity and war crimes committed in Sierra Leone since 1996.[24]
In *Prosecutor v. Sam Hinga Norman*[25] the defence made certain objections in a motion relating to the substantive jurisdiction of the Special Court. In essence, the defence asserted that the Special Court violates the principle of *nullum crimen sine lege* since the crimes mentioned in the Indictment were not part of customary international law at the relevant times, neither were they criminalized under Sierra Leonean criminal law at those times. The prosecutor replied to the defence and argued as follows:

> The principle of *nullum crimen sine lege* should not be rigidly applied to an act universally regarded as abhorrent. The question is whether it was foreseeable and accessible to a possible perpetrator that the conduct was punishable.[26]

The majority of the judges in this hearing favoured the approach submitted by the prosecutor, namely that the emphasis should be on the conduct, rather than on the specific description of the offence in substantive criminal law.
The examples from the Nuremberg, Tokyo and the Special Court for Sierra Leone notwithstanding, I would argue that the margin of difference between national and international understandings of the legality principle in criminal law is not fundamental. At any rate, the protection of the legality principle in the Rome Statute (as mentioned above) is quite clear and in line with most national systems that provide for a rather strict protection of the legality principle. The relevance of the more fluid application of *nullum crimen sine lege* is perhaps limited to crimes (and criminalization) under customary international

22 T. Taylor, 'The Nuremberg Trials', 55 *Columbia Law Review* 4, 1955, pp. 488-525, at p. 516.
23 *See* discussion in A. Cassese, *International Criminal Law*, Oxford, Oxford University Press, 2003, pp. 143-144.
24 Statute of the Special Court for Sierra Leone (2002), reproduced in C. van den Wyngaert, *International Criminal Law – A Collection of International and European Instruments* (3rd edn), 2005, p. 307.
25 *Prosecutor v. Sam Hinga Norman (Decision on preliminary motion based on lack of jurisdiction – child recruitment)* Case No. SCSL-2004-14-AR72 (E) 31 May [2004].
26 *Prosecutor v. Sam Hinga Norman*, para. 2.

criminal law. There is no reason why conventional international criminal law should be treated differently from statutory criminal law at the national level.[27]

3.3.3 *State Sovereignty and the Impact of International Criminal Justice*

It was noted above that the evolving system of international criminal justice can be viewed as a constitutional limitation on the freedom of sovereign states to conduct their affairs as they wish. It is trite, of course, that international law generally has this constitutional effect. While the Charter of the United Nations is based "on the principle of the sovereign equality of all its Members,"[28] other developments in international law, notably human rights, international criminal justice, and, indeed, environmental law, shape the content and scope of state sovereignty. In essence, it is safe to say that states can no longer act and do as they want – not in terms of their relations with other states and to a growing extent, internally as well.

Bruce Broomhall eloquently describes the impact of international (criminal) law on the notion of state sovereignty:

> The idea that sovereignty does not arise in a vacuum, but is constituted by the recognition of the international community, which makes its recognition conditional on certain standards, has become increasingly accepted in the fields of international law and international relations. Such limits are held always to have been imposed by the community on the recognition of its members, but to be subject to development over time. From this perspective, crimes under international law can be understood as a formal limit to a State's legitimate exercise of its sovereignty, and so in principle justify a range of international responses (subject to the rest of international law, including that relating to the use of force).[29]

I thus understand sovereignty to be a notion constantly changing and evolving in light of the growing importance of international law, including the content and institutions of international criminal law. International criminal law should, however, not be seen as a negative limitation of state sovereignty. It also empowers states to act as agents of a normative framework premised on the rule of law, and a drive to end impunity for the worst crimes affecting the whole of humankind. This brings me to the role of states vis-à-vis the ICC: the primary role-player in the modern international criminal justice system.

---

27 For more on this debate, *see* M. Swart, 'Legality as Inhibitor: The Special Place of nullum crimen sine lege in the Jurisprudence of International Criminal Tribunals', 30 *South African Yearbook of International Law*, 2005, pp. 33-49.

28 Art. 2(4) UN Charter.

29 B. Broomhall, *International Justice & The International Criminal Court – Between Sovereignty and the Rule of Law*, Oxford, Oxford University Press, 2003, p. 43.

3.3.4 *The International Criminal Court*

The creation of the ICC in terms of the Rome Statute of 1998[30] was not an End of History[31] moment; it did not represent an end point even in terms of the long historical quest for a system or structure to end impunity. It can be seen as an important development, even a starting point,[32] in the quest to end impunity for the worst crimes against humankind.[33] For present purposes a number of features of the ICC will be highlighted. The aim is to briefly identify and discuss those features of the ICC that are viewed as relevant for this contribution.

3.3.4.1 The Crimes Within the Substantive Jurisdiction of the ICC

The Core Crimes

The ICC has jurisdiction over the crimes of genocide, crimes against humanity, war crimes and the crime of aggression (although the latter crime is not yet within the effective jurisdiction of the Court).[34] The crimes listed were regarded by the drafters of the Rome Statute as the *most serious crimes of concern to the international community as a whole.*[35]

---

30 Rome Statute of the International Criminal Court, 1998), UN doc. A/CONF 183/9, *International Legal Materials* 1998, p. 999.

31 This refers to the much debated book by F. Fukuyama, *The End of History and the Last Man*, Penguin, London 1992. At the time, Fukuyama argued that, after the fall of the Berlin Wall and the collapse of the Soviet empire, the winner is clear: market economy and liberalism. It is plausible to see the progress of international criminal law, and the institutions of international criminal justice during the 1990s as part of the new world order made possible by the collapse of communism. It is true that the thaw in international relations after the end of the Cold War made possible the consensus in the Security Council, which in turn adopted the statutes of the first international criminal tribunals since Nuremberg and Tokyo, namely the International Criminal Tribunal for the Former Yugoslavia (1993) and the International Criminal Tribunal for Rwanda (1994). It is quite reasonable to see the adoption of the Rome Statute of the International Criminal Court in 1998 as a logical extension of the process that started in the early 1990s. However, the ICC, which purports to be a permanent international tribunal, cannot be seen as a culmination or end point. In many ways it is a starting point.

32 *See* observations by Philippe Kirsch (who was elected the first President of the ICC). P. Kirsch, 'Introduction', *in* H. von Hebel, J. Lammers & J. Schukking (Eds.), *Reflections on the International Criminal Court – Essays in Honour of Adriaan Bos*, The Hague, TMC Asser Press, 1999, p. 1.

33 For a discussion of the historical processes that culminated in the adoption of the Rome Statute of the ICC, as well as a broader historical perspective on the trends and currents in international criminal justice, *see* C. Bassiouni, *The Legislative History of the International Criminal Court*, Vol. I, New York, Transnational Publishers, 2005, pp. 3-121.

34 The Resolution on the Crime of Aggression, adopted at the Kampala Review Conference of the International Criminal Court (2010) provides for a definition of aggression and for conditions for the exercise of ICC jurisdiction over the crime of aggression. It is not yet in operation. At the time of writing, only five states have ratified the resolution. The ICC may exercise jurisdiction over the crime of aggression once thirty states have ratified the amendments on aggression, and subject to a decision to be taken after 1 January 2017 by the states party to the Rome Statute of the ICC.

35 Art. 5(1) Rome Statute of the International Criminal Court.

This list of crimes can be viewed as at least containing the most serious crimes affecting humankind, but not necessarily complete, as I will discuss later on. The states that adopted the Rome Statute discussed various crimes to be included within the ICC's jurisdiction. The inclusion of the crime of genocide (based on the Genocide Convention of 1948) was not controversial. A bit more contentious were the definitions and scope of crimes against humanity and war crimes. After a considerable debate, the crime of aggression was also included, but on condition that a suitable definition must be drafted and conditions for the exercise of jurisdiction by the ICC of this crime be agreed upon. This indeed happened at the Kampala Review Conference of the ICC, which was held in 2010. However, the definition of the crime of aggression and the related provisions on this crime can at the earliest enter into force in 2017.[36]

Other Crimes Like Drug Trafficking and Terrorism

Apart from the above so-called core crimes under international law, the Rome Diplomatic Conference considered the inclusion of other crimes within the jurisdiction of the ICC. The most notable of these crimes were drug trafficking, terrorism, and violations of the Convention on the Safety of United Nations and Associated Personnel. A majority of states at the Rome Diplomatic Conference opposed inclusion of these treaty crimes.[37] The (rather unconvincing) rationale for the exclusion was that there already existed effective enforcement and cooperation regimes for these crimes.

Jerry Fowler, legal counsel for the Lawyers Committee for Human Rights and participant at the Rome Conference, pointed out that the inclusion of drug trafficking and terrorism enjoyed "significant support," but not as much as aggression. While drug trafficking and terrorism certainly affect the international community as well, there was simply not enough agreement on the inclusion of these crimes. Of course, the contentious scope and content of, to a lesser extent, drug trafficking and, to a large extent, terrorism, made them less obvious to include as the worst crimes affecting the whole of humankind.

What about Crimes against the Environment?

It was noted above that the ICC was established as a result of many legal and geopolitical factors that aligned in the 1990s. I also pointed out that the creation of the ICC should not be seen as an end point, but rather as an important starting point, or new chapter, in international criminal justice.

36 For a comprehensive discussion of the work of the Special Working Group on the Crime of Aggression, *see* Kemp, 2010, pp. 208-237. For a discussion of the definition of aggression and the conditions for the exercise of jurisdiction by the ICC over this crime, as adopted at the Kampala Review Conference, *see* C. Kress & L. von Holtzendorff, 'The Kampala Compromise on the Crime of Aggression', 8 *Journal of International Criminal Justice*, 2010, pp. 1179-1217.

37 R.S. Lee (Ed.), *The International Criminal Court – The Making of the Rome Statute*, The Hague, Kluwer Law International, 1999, p. 81.

While the outcome of the Rome Diplomatic Conference on the ICC was, on balance, satisfactory in terms of the crimes included in the jurisdiction of the ICC, activists and environmental lawyers can rightly point to the exclusion of crimes against the environment as an important omission, hopefully to be rectified at a future Review Conference. The exclusion of crimes against the environment was by no means the result of bad preparatory work. In 1996, two years prior to the Rome Diplomatic Conference, a document on crimes against the environment was published in the context of the work by the International Law Commission on the Draft Code of Crimes against the Peace and Security of Mankind – including the draft statute for an ICC.[38] The proposed crimes against the environment were never adopted at Rome. However, the document on crimes against the environment can still serve as a very useful starting point for a debate on the inclusion of crimes against the environment (and even more broadly, crimes in the context of climate change) at a future Review Conference. Some of the important aspects of the document on crimes against the environment are discussed below.

3.3.4.2 The Legal Nature of the Rome Statute of the ICC

The Rome Statute of the ICC is a multilateral treaty that is binding on the states party to it. While the status of the Rome Statute under general public international law seems to be clear, there are different views as to the exact legal nature of the Rome Statute, and – important from a criminal law perspective – the question whether it (also) binds individuals (and not only states). The *first view* is that the Rome Statute is jurisdictional in nature only. This means that the provisions that deal with international crimes (war crimes, genocide, crimes against humanity and aggression) are addressed to the ICC itself. Thus, these provisions do not bind individuals; the provisions only set out the scope of the ICC's subject matter jurisdiction. In terms of this view, the source of substantive norms of criminal law pertaining to individuals would be in customary international law. The *second proposition* is that the Rome Statute is essentially substantive in nature. It binds individuals by creating substantive norms with criminal sanctions. Marko Milanovic framed the proposition as follows:

> If the [Rome] Statute was to be seen as being substantive in nature, then it would be *creating* the international crimes defined within it quite independently of the status of these crimes in customary law. Customary law would be relevant only to the extent that the provisions of the Statute might refer to it, and charges before the Court could generally not be challenged on the basis that they do not conform to custom. Indeed, the Statute could create

38 Document ILC (XLVIII)/DC/CRD.3 Document on crimes against the environment, prepared by Mr Christian Tomuschat, member of the International Law Commission, C. Tomuschat, *Yearbook of the International Law Commission*, Vol. II(1), 1996, available at <www.un.org/law/ilc/index.htm> (last accessed 14 October 2012).

> and contain crimes which clearly do *not* exist in customary international law. Think, for example, of the various proposals to include international drug trafficking or even ecocide within the Statute [. . .][39]

Whether the Rome Statute is primarily jurisdictional in nature or primarily substantive in nature is open to debate. What is clear is that the answer to the question is perhaps less problematic when one is dealing with the current crimes within the ICC's jurisdiction – war crimes, crimes against humanity and genocide. These crimes have a firm basis in customary international law, and one could therefore argue that individuals are also bound by these criminal norms. The Rome Statute simply codified what already existed under customary international law, thus confirming and enforcing universal norms. The jurisdictional and the substantive roles of the Rome Statute thus overlap to a large extent.

Crimes that do not have a basis in customary international law might pose more problems from a legality point of view. Crimes against the environment do not form part of customary international law. There is also at present no international treaty that provides for international crimes against the environment, or for crimes in the context of climate change. This means that even if the Assembly of states party to the Rome Statute would one day adopt a definition of 'ecocide' or of crimes against the environment or of crimes in the context of climate change, legal problems might still arise – no less as a result of the different views of the legal nature of the Rome Statute. Such uncertainty will inevitably also feed into questions about legality. Nevertheless, in the absence of custom, the only viable way forward would be for the adoption of a definition to be included in the Rome Statute. This possibility is further explored under Part 4 below.

3.3.4.3 ICC Has Jurisdiction Only over Natural Persons (Not Legal Entities)

Article 25(1) of the Rome Statute of the ICC provides as follows: "The Court shall have jurisdiction over natural persons pursuant to this Statute." There is no provision in the statute for corporate or state criminal liability.

There was considerable debate at the Rome Diplomatic Conference on whether to include corporate criminal liability. France proposed the recognition and inclusion of corporate criminal liability on the following grounds: First, the French concept of criminal liability for *personnes morales* is an established criminal law notion. Second, there is a moral imperative to punish all entities (natural or legal) who are responsible for the worst crimes affecting the whole of humankind. Third, corporate criminal liability presented the ICC with a practical mechanism for ensuring compensation – especially since individual perpetrators on their own might not have the resources to fund adequate compensation for victims. Fourth, the

39 M. Milanovic, 'Aggression and Legality – Custom in Kampala', 10 *Journal of International Criminal Justice*, 2012, p. 172. Emphasis in the original.

notion of corporate criminal liability can serve as a deterrent and can foster a culture of caution in the context of profit-motivated decision-making processes.[40]

The notion of corporate criminal liability is well established in many national legal systems. There are also notable exceptions, for instance Germany. Although the German Criminal Code provides for criminal liability of certain administrators and officers of corporations, there is no general principle of liability for the legal entity, the corporation.[41]

Despite the relative common acceptance of corporate criminal liability at the national level, the debate at the Rome Diplomatic Conference resulted in a rejection of the French proposal on the inclusion of corporate criminal liability under the Rome Statute. The rejection of the proposal was not primarily motivated by a lack of understanding that corporations are important role-players and actors in terms of potential liability for gross human rights violations and atrocities. Indeed, many delegations at the Rome Conference noted the fact that many conflicts and instances of gross human rights violations were in no small measure aided and abetted by corporations. These include the involvement of media entities in the Rwandan genocide, the forcible removal and transfer of people as a result of the activities of multinational oil companies, and, decades earlier, the involvement of corporations (including big banks) in the Holocaust.[42]

Because of the workings of the system of complementarity, which is discussed below, some delegations at the Rome Conference felt that the inclusion of corporate criminal liability would cause jurisdictional and practical problems. This means that where national legal systems do not recognize corporate criminal liability, this could hardly be seen as their being unwilling or unable to deal with the crime in question, as per the complementarity regime. Because of the difficulties mentioned above, France ultimately decided to withdraw the proposal.

The exclusion of corporate criminal liability from the ICC jurisdictional regime is relevant for this contribution, since it is assumed that corporations are not only active role-players in the global economy, but indeed also responsible for the activities that cause climate change. The implications of this observation are further explored below.

3.3.4.4 The Principle of Complementarity

The jurisdictional regime of the ICC can perhaps best be described as justice in reserve: the ICC will step in only if states party to the Rome Statute are unwilling or unable to investigate and prosecute the crimes that fall under the jurisdiction of the ICC. This is

40 D. Stoitchkova, *Towards Corporate Liability in International Criminal Law*, Antwerp, Intersentia, 2010, p. 14.

41 For a comparative perspective, *see* G. Kemp (Ed.), *Criminal Law in South Africa*, Cape Town, Oxford University Press, 2012, pp. 215-218.

42 Stoitchkova, 2010, p. 15. For more detail *see* S.D. Bachmann, *Civil Responsibility for Gross Human Rights Violations – The Need for a Global Instrument*, Pretoria, Pretoria University Press, 2007.

known as the principle of complementarity. The principle is provided for in Article 17 of the Rome Statute:

> 1. having regard to paragraph 10 of the Preamble and article 1, the Court shall determine that a case is inadmissible where:
>   (a) The case is being investigated or prosecuted by a State which has jurisdiction over it, unless the State is unwilling or unable genuinely to carry out the investigation or prosecution;
>   (b) The case has been investigated by a State which has jurisdiction over it and the State has decided not to prosecute the person concerned, unless the decision resulted from the unwillingness or inability of the State genuinely to prosecute;
>   (c) The person concerned has already been tried for conduct which is the subject of the complaint, and a trial by the Court is not permitted under article 20, paragraph 3;
>   (d) The case is not of sufficient gravity to justify further action by the Court.
> 2. In order to determine unwillingness in a particular case, the Court shall consider, having regard to the principles of due process recognized by international law, whether one or more of the following exist, as applicable:
>   (a) The proceedings were or are being undertaken or the national decision was made for the purpose of shielding the person concerned from criminal responsibility for crimes within the jurisdiction of the Court referred to in article 5 [currently genocide, crimes against humanity and war crimes];
>   (b) There has been an unjustified delay in the proceedings which in the circumstances is inconsistent with an intent to bring the person concerned to justice;
>   (c) The proceedings were not or are not being conducted independently or impartially, and they were or are being conducted in a manner which, in the circumstances, is inconsistent with an intent to bring the person concerned to justice.
> 3. In order to determine inability in a particular case, the Court shall consider whether, due to a total or substantial collapse or unavailability of its national judicial system, the State is unable to obtain the accused or the necessary evidence and testimony or otherwise unable to carry out its proceedings.

It is obvious from the text of Article 17 that the issue of complementarity is very complex.[43] It involves the intricate relationship between the ICC and states. It underscores the ICC as an instrument of international criminal justice in reserve. Complementarity puts the emphasis on the *national* application of international criminal law. At the same time, one can view the principle of complementarity as a form of fairness and quality control over national criminal justice systems. In this sense, it reminds us of the constitutional role of international criminal justice mechanisms such as the Rome Statute of the ICC, as noted above. Thus, one can say that the principle of complementarity is also a manifestation of global governance.

3.4 The Potential Role of International Criminal Justice in Environmental and Climate Change Law

From the above discussion, it should be clear that the evolving system of international criminal justice (both in terms of substantive law and in terms of enforcement) can be regarded as an important piece of global governance. The ICC as the 'face' of international criminal justice is imperfect. Not only is it not meant to be the sole or even the most important enforcer of international criminal law, it also lacks certain mechanisms that would be necessary to address possible crimes against the environment and, more broadly, climate change.

The exclusion of corporate criminal liability, complementarity, and the lack of substantive jurisdiction over crimes against the environment were noted. But it was also noted that the system of international criminal justice, including the ICC, is an evolving system. It is dynamic. The Rome Statute of the ICC also provides for amendment procedures. The mechanics and technical aspects of amendment aside, I will now turn to discuss the possibility to criminalize crimes against the environment (including crimes of sufficient gravity to be regarded as crimes that contribute to climate change). I will then briefly discuss the ICC as a possible forum for the prosecution of crimes against the environment.

3.4.1 Attempts to Criminalize Crimes against the Environment under International Law

I will address the possible criminalization of crimes against the environment in terms of two distinct but related subject matters: environmental crimes (or crimes against the environment) and crimes that contribute to climate change.

43 For a discussion *see* A. Zahar & G. Sluiter, *International Criminal Law*, Oxford, Oxford University Press 2008, pp. 455-456.

It is beyond the scope of this exploratory contribution to fully analyse the elements of crimes against the environment as a crime under international law. The aim here is to outline the most important aspects and basic elements to the extent that they are relevant for purposes of a discussion about the general theme of international criminal justice and climate change.

3.4.1.1 Crimes against the Environment

It was noted above that the International Law Commission prepared a document on Crimes against the Environment in the context of the preparatory work that included a draft statute for an ICC. While the ICC eventually materialized (in terms of the Rome Statute of the ICC), crimes against the environment were not part of the package.

The short history of the Document on Crimes against the Environment (henceforth the 'DCAE') highlights the fact that the normative underpinnings of international criminal law always (or at least since Nuremberg) seemed to have favoured the *protection of the human person directly*. Indeed, Tomuschat pointed out that at Nuremberg, no charges were brought on account of the immense damages to the natural environment during World War II. Even the ban on "poison or poisoned weapons" in terms of the so-called Law of The Hague on the Rules and Methods of Warfare sought to prevent the immediate impact on military personnel, not the natural environment – even though these weapons could have lasting damaging effects on the environment.[44]

The lack of substantive and enforcement recognition of crimes against the environment at Nuremberg resulted in the fact that the International Law Commission, which was tasked to codify the principles of Nuremberg after the completion of the Nuremberg Trials, also did not address the issue of crimes against the environment. The Nuremberg Principles are therefore silent on the matter, and the subsequent first Draft Code of Offences against the Peace and Security of Mankind (1954) also did not include any new crimes that were not mentioned in the Nuremberg Principles.

Decades later, in 1986, however, environmental awareness and the work of a special rapporteur resulted in the submission that the list of crimes against humanity should include a category of crimes against the environment. This form of crime against humanity would be: *any serious breach of an international obligation of essential importance for the safeguarding and preservation of the human environment.*[45]

During the debates at the International Law Commission's 38th session, the majority of speakers supported the notion of crimes against the environment (as crimes against

44 Document ILC (XLVIII)/DC/CRD.3 Document on crimes against the environment, prepared by Mr Christian Tomuschat, member of the International Law Commission, Tomuschat, 1996, para. 3.

45 *Yearbook of the International law Commission*, 1986, Vol. II (Part One), pp. 85-86, cited in Document ILC (XLVIII)/DC/CRD.3 Document on crimes against the environment, prepared by Mr Christian Tomuschat, member of the International Law Commission, Tomuschat, 1996, para. 5.

humanity). Some offered caution on the basis that crimes against the environment do not fit within the structure and normative foundations of crimes against humanity. Others warned that more clarity is needed (also important from a legality point of view). In terms of the general principles of criminal liability, it was noted that intent (*dolus*) will be required for liability for crimes against the environment. Following the debate, and following the further work of the Special Rapporteur, a revised article on crimes affecting the environment made it into the seventh report (that preceded the adoption of the 1991 Draft Code of Crimes against the Peace and Security of Mankind). The definition of crimes affecting the environment in the seventh report (1989) thus provides that the following constitute crimes against humanity: *Any serious and intentional harm to a vital human asset, such as the human environment.*[46]

In the 1991 Draft Code of Crimes against the Peace and Security of Mankind, crimes against the environment were eventually not listed as crimes against humanity, but rather as a new crime, namely, causing wilful and severe damage to the environment: *An individual who wilfully causes or orders the causing of widespread, long-term and severe damage to the natural environment shall, on conviction thereof, be sentenced. (to . . .).*[47] It should be mentioned that this crime could be committed in times of peace or during an armed conflict. It was not supposed to be linked to an armed conflict. However, by 1996 the independent crime of causing wilful and severe damage to the environment was not listed in the Draft Code. Article 20 of the Draft Code of Crimes against the Peace and Security of Mankind (1996)[48] provided for a war crime in the form of the use of methods or means of warfare not justified by military necessity with the intent to cause widespread, long-term and severe damage to the natural environment and thereby gravely prejudicing the health or survival of the population. The notion of crimes against the environment thus moved from a form of crimes against humanity (the reports preceding the 1991 Draft Code) to an independent crime of causing wilful and severe damage to the environment (1991 Draft Code), and back to a more limited notion of crimes against the environment as a war crime during an armed conflict (1996 Draft Code).

As noted above, none of the modalities for the criminalization of crimes against the environment (as an independent group of crimes under international law) were eventually adopted at the Rome Diplomatic Conference on the ICC in 1998. The Rome Statute of the ICC provides, in line with the approach adopted for purposes of the 1996 Draft Code, for

46 *Yearbook of the International law Commission*, 1989, Vol. II (Part One), p. 86, cited in Document ILC (XLVIII)/DC/CRD.3 Document on crimes against the environment, prepared by Mr Christian Tomuschat, member of the International Law Commission, Tomuschat, 1996, para. 7.

47 Art. 26 Draft Code of Crimes against the Peace and Security of Mankind, 1991, *Report of the International Law Commission*, 43d Session, UNGAOR, 46th Session, Supp. No. 10, A/46/10 (1991), reproduced in Van den Wyngaert, 2005, pp. 323-329.

48 Draft Code of Crimes against the Peace and Security of Mankind, 1996, *Yearbook of the International Law Commission*, 1996, Vol. II(2) available at <http://www.un.org/law/ilc/texts/dcodefra.htm> (last accessed 14 October 2012).

the war crime (in the context of an international armed conflict) of intentionally launching an attack in the knowledge that such attack will cause incidental loss of life or injury to civilians or damage to civilian objects or *widespread, long-term and severe damage to the natural environment which would be clearly excessive in relation to the concrete and direct overall military advantage anticipated.*[49]

It is thus clear that the Rome Statute of the ICC is much more restricted in terms of the substantive notion of crimes against the environment. It is a form of war crime in the context of an international armed conflict. It is a far cry from the more progressive view of crimes against the environment as either a stand-alone international crime or as a crime against humanity – committed during times of peace or during an armed conflict.

Next, I will briefly consider the possibility of reviving the debate about crimes against the environment for inclusion in the Rome Statute (that is during a future revision conference), or of providing for crimes against the environment in a separate convention (comparable to the Genocide Convention,[50] or the four Geneva Conventions[51]).

3.4.1.2 A More Comprehensive Criminalization of Crimes against the Environment (Incorporating the Dangers of Climate Change)

Any discussion of the possible criminalization of crimes against the environment, incorporating the element of climate change, must take due notice of the normative context and developments of the past twenty years. The impetus was provided by the adoption of the United Nations Framework Convention on Climate Change (1992).[52] The primary aim of the Framework Convention is the stabilization of greenhouse gas (GHG) concentrations in the atmosphere at a level that would prevent dangerous anthropogenic interference with the climate system. Such a level should be achieved within a time frame sufficient to allow ecosystems to adapt naturally to climate change, to ensure that food production is not threatened and to enable economic development to proceed in a sustainable manner.[53]

The very broad normative and policy objectives reflected in the Framework Convention can at best be the starting point for a discussion on the elements of a future international crime against the environment. This is even more so if one wants to criminalize the acts

49 Art. 8(2)(b)(iv) Rome Statute of the International Criminal Court, 1998.

50 Convention on the Prevention and Suppression of the Crime of Genocide, 1948, *UN Treaty Series*, Vol. 78, p. 227.

51 Convention for the Amelioration of the Condition of the Wounded and Sick in Armed Forces in the Field, 1949) *UN Treaty Series*, Vol. 75, p. 31; Convention for the Amelioration of the Condition of the Wounded, Sick and Shipwrecked Members of the Armed Forces at Sea, 1949, *UN Treaty Series*, Vol. 75, p. 85; Convention relative to the treatment of Prisoners of War, 1949, *UN Treaty Series*, Vol. 75, p. 135; Convention relative to the Protection of Civilian Persons in Time of War, 1949, *UN Treaty Series*, Vol. 75, p. 287.

52 United Nations Framework Convention on Climate Change, 1992, 31 *ILM* 849. For commentary on the subject of climate change and the normative impact of the Framework Convention, *see* J.S. Dryzek, R. Norgaard & D. Schlosberg (Eds.), *The Oxford Handbook of Climate Change and Society*, Oxford, Oxford University Press, 2011; J. Vinuales, 'Balancing Effectiveness and Fairness in the Redesign of the Climate Change Regime', 24 *Leiden Journal of International Law*, 2011, p. 223.

53 Art. 2 of the Framework Convention.

that cause or harmfully contribute to climate change. With reference to the crime of causing harm to the environment, as provided for in the 1991 Draft Code (discussed above), the Government of the United States, for instance, reacted as follows:

> This article, dealing with damage to the environment, is perhaps the vaguest of all the articles. The article fails to define its broad terms. There is no definition of "widespread, long-term and severe damage to the natural environment." Similarly, the term "wilfully" is not defined, thereby creating considerable confusion concerning the precise volitional state needed for the imposition of criminal liability. The term "wilfully" could simply mean that the defendant performed an act voluntarily, i.e. without coercion, that had the unintended effect of causing harm to the environment. "Wilfully" could also be construed to impose criminal liability only when the defendant acted for bad purpose, knowing and intending to cause serious harm to the environment. As presently drafted, the meaning of "wilfully" is subject to a variety of interpretations. This confusion is magnified by the draft Code's failure throughout to specify the necessary mental and volitional states needed for the imposition of criminal liability.

The above reaction is indicative of an approach highlighting the strict application of general principles of criminal law. Aspects like the legality principle (which was earlier discussed as one of the key features of the international criminal justice system) demarcate and limit the criminal law content of broader normative (policy-oriented) propositions. A number of considerations in the context of the criminalization of crimes against the environment should be considered:

The Protected Interest: The Environment

It is clear from the debates, reports and various draft texts and other international instruments informing the various Draft Codes of Crimes against Peace and Security of Mankind that 'environment' connotes 'human environment' as well as 'natural environment'. The fifteenth International Congress of Penal Law, which was held in Rio de Janeiro in 1994, adopted the following description of environment in its resolution on crimes against the environment:

> Environment means all components of the earth, both abiotic and biotic, and includes air and all layers of the atmosphere, water, land, including soil and mineral resources, flora and fauna, and all ecological inter-relations among these components.[54]

54 Reproduced in *Yearbook of the International law Commission*, 1986, Vol. II (Part One), pp. 85-86, cited in Document ILC (XLVIII)/DC/CRD.3, Document on crimes against the environment, prepared by Tomuschat, 1996, para. 25.

Thus, *natural* environment is the emphasis. Other aspects of the 'environment' in which we live (including our cultural spaces) are best protected under a separate framework, as is indeed the case with war crimes, which criminalizes wanton destruction of cultural property.

Gravity and Scale: Requirements of Seriousness

Crimes under international law, especially the so-called atrocity crimes (genocide and crimes against humanity) and the crime of aggression, normally become relevant for the international criminal justice system because of their gravity and scale. Thus, for instance, the Rome Statute of the ICC proclaims that it has jurisdiction over persons for the most serious crimes of international concern (Art. 1), and further, that the ICC shall determine that a case is inadmissible where it is not of sufficient gravity to justify further action by the Court (Art. 17).

To the extent that one can say that climate change (caused by the increase in GHG concentration in the atmosphere)[55] affects the environment in such a way as to cause, for instance, wholescale destruction of ecosystems, it must be viewed as serious and of sufficient gravity for purposes of international criminal justice action.

The requirement of seriousness for purposes of international criminal justice is in line with international instruments that criminalize crimes under international law. Article 55 of Additional Protocol I to the Geneva Conventions (1949),[56] for instance, refers to the protection of the natural environment against widespread, long-term and severe damage.

Harm to the Environment: An International Dimension

The type of crimes relevant for international criminal justice are crimes that affect the whole of humankind. This is a normative perspective. It is not a factual question of territorial impact. International in this context does not mean "more than one jurisdiction" or "more than one state." The inherent nature of crimes like genocide and crimes against humanity is that they affect and shock the whole of humanity; hence the term *international* crime.

It has been suggested that an international element is present in a crime when the *commons of humankind have been affected*.[57] The environment, climate and complex ecosystems are, to my mind, commons of humankind and any severe or serious damage or long-term damage to the environment (including as a result of climate change) is, at least as a matter of principle, worthy of international action, including via the system of international criminal justice.

55 Art. 2 Framework Convention on Climate Change.

56 Protocol Additional to the Geneva Conventions of 12 August 1949, and relating to the Protection of Victims of International Armed Conflicts (1977) (Protocol I), UN Doc. A/32/144, 15 August 1977, reproduced in Van den Wyngaert, 2005, pp. 455-494.

57 Tomuschat, 1996, para. 32.

An Autonomous Crime against the Environment (Not Linked to Armed Conflict or Crimes against Humanity)

It was noted above that, at present, the environment is a protected aspect for purposes of war crimes law. It was not adopted as a crime against humanity or as an autonomous crime under the Rome Statute of the ICC, or any other international criminal tribunal. There is, at present, not a comprehensive treaty criminalizing crimes against the environment.

The aim here is not to provide the content of a comprehensive instrument on crimes against the environment. It must be noted that there are certain pointers in this regard. Some of them are briefly mentioned here.

First, attempts to codify and provide for a comprehensive framework on the criminalization of crimes against the environment (or 'ecocide') are not new. In the 1970s a draft International Convention on the Crime of Ecocide was prepared by the international law scholar Richard Falk. This draft convention was proposed and discussed in the context of an evaluation of the effectiveness of the Genocide Convention of 1948.[58] The spirit of the draft text is today kept alive by the Ecocide Project at the Human Rights Consortium, School of Advanced Study, University of London.[59] The project's concept of ecocide rests on two key notions:

> *Ascertainable ecocide* is a crime of consequence, primarily arising out of corporate damaging and destructive activity which is primarily governed by ineffectual and nominal civil legislation. In the case of *non-ascertainable ecocide* (other causes – *e.g.* tsunami, rising sea levels – climate change driven) there is currently no coherent international mechanism in place to help territories that are rendered unable to self-govern and are in need of emergency assistance. Instead, we deal with each disaster on a 'case by case' basis after the event.

The concept of ecocide, together with the work that was done by the International Law Commission in the context of the Draft Codes of Crimes against Peace and Security of Mankind, provides a solid basis for further research and proposals on a comprehensive international criminal justice response to crimes against the environment, including the consequences of climate change.

To criminalize an autonomous crime against the environment is an important (indeed, vital) step, but one must also take into consideration the necessary enforcement regime. A few points in this regard are highlighted below.

---

58 R.A. Falk, 'Environmental Warfare and Ecocide – Facts, Appraisal, and Proposals', *in* Thee & Marek, 1 *Bulletin of Peace Proposals*, Oslo/Bergen, Universitersforlaget, 1973, pp. 80-96.

59 <www.sas.ac.uk/hrc/projects/ecocide-project> (last accessed 14 October 2012).

3.4.2 *A Framework for Enforcement*

3.4.2.1 National Level: States as Agents of the International Community

Serious crimes against the environment affect the whole of humankind. A global (or, in legal terms, *international*) response is required. It is not enough or adequate to leave enforcement to individual states. It is an open question whether states are always willing or able to prosecute violations of environmental laws. South Africa is a case in point. A 2012 report[60] noted that, while the National Environmental Management Act[61] provides for a number of criminal offences, individuals and companies often apply to the relevant ministry for *ex post facto* authorization of relevant parts of the law that were violated. The Act does indeed provide for such *ex post facto* authorization, but an almost institutionalized practice of this nature seems to be an abuse and violation of the letter and spirit of the Act. It reflects poorly on the political and institutional will to combat violations of environmental laws. It furthermore underscores the point that states alone cannot be trusted with matters affecting the environment – especially if the impact goes beyond the local environment. An international or transnational approach is needed. The eventual criminalization of crimes against the environment at the international level might help to force states to take seriously their custodial duties as protectors of the environment on behalf of all humanity.

3.4.2.2 Regional Level

In terms of geography and political pragmatism, it might make sense to advance an agenda of criminalization of crimes against the environment at the regional level. This might, at least in the short or medium term, be more realistic than to push for a true international regime on criminal liability for crimes against the environment.

The Council of Europe Convention for the Protection of Environment through Criminal Law[62] does not create a direct or vertical supranational or international enforcement regime. It rather focusses on the creation of certain obligations (contracting states must, for instance, introduce certain specific criminal provisions into their domestic criminal laws) and (horizontal) international cooperation in criminal matters.

In the context of the European Union, a number of so-called eurocrimes have developed. A number of EU instruments provide for the criminalization of certain protected interests. These include crimes such as crimes against fair competition, crimes against the integrity of the financial sector, crimes against human dignity, crimes against the democratic

60 <www.legalbrief.co.za/> (last accessed 11 October 2012).

61 Act 107 of 1998.

62 CETS No. 172. available at <http://conventions.coe.int/Treaty/en/Html/172.htm> (last accessed 15 October 2012).

society and crimes against the environment. In terms of EU Directives,[63] Member states must criminalize various forms of conduct related to the environment.

The Draft Protocol on Amendments to the Protocol on the Statute of the African Court of Justice and Human Rights (African Union) provides for the creation of an African Court with international criminal law jurisdiction over certain crimes of regional and international concern. These include the classic 'core' crimes under international law (genocide, war crimes, crimes against humanity) as well as other crimes of concern such as drug trafficking, human trafficking and corruption. It also provides that the Court shall have jurisdiction over the crime of trafficking in hazardous wastes. This is the closest that the Draft Protocol gets to the criminalization of crimes against the environment.

At first glance, the above regional examples (Council of Europe, EU and AU) seem to be steps in the right direction, namely a regional approach to crimes that clearly affect more than just national or parochial interests.

It is beyond the scope of this contribution to fully analyse these regional instruments or proposed instruments. Regarding developments in the AU, the following can be noted. The creation of a criminal chamber in the African Court of Justice and Human Rights, with jurisdiction over persons responsible for serious international crimes or crimes that affect the African region as a whole, should in principle be a commendable development to end impunity for these crimes. The problem is that many observers see the creation of this 'African Criminal Court' as a political stunt aimed at undermining or disrupting the jurisdictional regime of the ICC.[64] More than thirty African states are also states party to the Rome Statute of the ICC. It is thus not clear how the proposed African Criminal Court will fit into the complementarity regime of the ICC. Since the ICC does not have jurisdiction over crimes such as drug trafficking, corruption and the (limited) crime against the environment of trafficking in hazardous waste, the African Criminal Court may yet prove to be a useful supplementary criminal regime to the ICC. A possible consequence might be that a (relatively) successful regional criminal regime can serve as impetus for an expanded substantive jurisdiction of the ICC over crimes such as human trafficking and environmental crimes. Of course, it could also lead to nothing. Only time will tell.

---

63 For instance Directive 2008/99. For a discussion *see* A. Klip, *European Criminal Law – An Integrative Approach* (2nd edn), Antwerp, Intersentia, 2012, p. 218.

64 P. Manirakiza, 'Towards and African Criminal Court: Contribution or obstruction to International Criminal Justice?, unpublished paper presented at the international conference: *Africa and the Future of International Criminal Justice*, University of the Witwatersrand, Johannesburg, 14-16 July 2010.

3.4.2.3 International Level

The ICC is the best available vehicle for the direct application of international criminal law at the international level. It is, of course, complementary to national criminal justice systems. The emphasis is thus on the national application and enforcement of international criminal law. The ICC has limitations as well. It does not have jurisdiction over legal persons (corporations). Its substantive jurisdiction is, at present, limited to crimes against humanity, war crimes and genocide. The Rome Statute of the ICC provides for amendment procedures, and the first Review Conference on the Rome Statute (held at Kampala in 2010) proved that it is indeed possible to expand the substantive jurisdiction of the ICC. However, the inclusion of the crime of aggression, long recognized as one of the core crimes under international law and indeed described as the "supreme international crime" at Nuremberg, was perhaps less difficult in the end precisely because of the historical basis for the criminalization of aggression. Crimes against the environment, let alone 'climate change', are yet to be regarded as universally recognized criminal notions.

It is not only at the substantive level that environmental crimes/climate change might prove to be problematic in terms of the ICC regime. As pointed out above, the ICC has jurisdiction only over natural persons. Any expansion of the substantive jurisdiction of the ICC to include crimes against the environment of conduct that causes climate change must also be accompanied by an expanded jurisdiction of the ICC to include corporations.

3.5 CONCLUDING REMARKS

If we accept, as we should, that climate change affects the whole of humankind, then we also need to accept that a global strategy to address this phenomenon is needed.

International criminal law, and the evolving system of international criminal justice, has the potential to be a key 'piece' of global governance. Climate change and environmental changes can be addressed only in terms of a global response. States on their own, with territorial or parochial approaches to law enforcement (including criminalization of crimes against the environment) cannot effectively address the challenges of climate change. Regional approaches seem to underscore the national implementation of environmental laws, but do not present a real transnational enforcement regime. It is therefore argued that the international criminal justice system, with the ICC as a key role-player, has the potential to be an important legal response to a changing environment. Much work needs to be done in terms of the substantive and enforcement jurisdiction of the ICC. As was the case with the crime of aggression, the process to provide for crimes against the environment (or crimes in the context of climate change) could potentially rely on considerable groundwork that has already been done. The structure of the Rome Statute is there. Questions about legality remain (and will always be a concern), but that does not mean that the ICC should be out of reach as a potential role-player in the ultimate fight against climate change that affects the whole of humankind.

The transformational impact of amendments to the Rome Statute of the ICC in the context of crimes that affect the environment and crimes in the context of climate change will be that many states will be able to act as agents of a truly global response to climate change via the complementarity enforcement regime provided for in the Rome Statute of the ICC. (International) criminal justice might yet prove to be a valuable part of the package of legal responses to climate change.

4 CLIMATE CHANGE LAW AND POLICY IN THE AFRICAN UNION AND SELECTED AFRICAN COUNTRIES

*Oliver C. Ruppel**

4.1 INTRODUCTION

Despite Africa's relatively low contribution to the world's total greenhouse gas (GHG) emissions, it is one of the most vulnerable continents to climate change.[1] Africa emits far less carbon than other continents. Africa's carbon dioxide (CO_2) emissions per year represent only a small fraction, 3.6%, of global emissions, yet 14% of the population of the world lives here.[2] The African continent accounts for only 3% of world energy consumption, and the average energy consumption of an African inhabitant is six times less than that recorded for individual inhabitants across the populations of the world. About 600 million people in Africa currently lack access to electricity.[3]

In the same light, Africa is particularly vulnerable as a consequence of a combination of stresses, and especially owing to poverty. The complexity of climate change will require the involvement of a diverse range of institutions.[4] It is expected that climate change will generate varied and significant impacts on national, regional and global economies; and it is also not unlikely that this will result in increased local and international conflict.[5]

Climate change poses an enormous threat to Africa's economic growth (through its harmful effects on natural systems and resources), long-term prosperity and the survival of

* Professor of Law, Faculty of Law, University of Stellenbosch, South Africa; and Coordinating Lead Author (CLA) of the Chapter on Africa in the forthcoming Assessment Report (AR5) of the Intergovernmental Panel on Climate Change (IPCC), Working Group II; ruppel@sun.ac.za.

1 M. Boko *et al.*, 'Africa', *in* M.L. Parry *et al.* (Eds.), *Climate Change 2007: Impacts, Adaptation and Vulnerability. Contribution of Working Group II to the Fourth Assessment Report of the Intergovernmental Panel on Climate Change*, Cambridge, Cambridge University Press 2007, pp. 433-467.

2 *See* <www.grida.no/graphicslib/detail/emissions-of-carbon-dioxide-in-africa-and-selected-oecd-countries_1400> (last accessed 17 January 2013).

3 *See* IRENA/International Renewable Energy Agency, *Prospects for the African Power Sector*, 2011, available at <www.irena.org/DocumentDownloads/Publications/Prospects_for_the_African_Power_Sector_abstract.pdf> (last accessed 17 January 2013).

4 R.O. Keohane & D.G. Victor, 'The Regime Complex for Climate Change', 9 *Perspective on Politics* 1, 2011, pp. 7-23.

5 W. Scholtz, 'The Promotion of Regional Environmental Security and Africa's Common Position on Climate Change', 10 *African Human Rights Law Journal*, 2010, pp. 1-25.

its already vulnerable populations. Climate change, variability and associated increased disaster risks are an additional burden to sustainable development in Africa, as well as a threat and impediment to achieving the Millennium Development Goals.[6]

In 2011, the South African city of Durban was in the international limelight as the host of the global climate negotiations.[7] The goal of these discussions was to advance the implementation of the United Nations Framework Convention on Climate Change (UNFCCC) and the Kyoto Protocol, and to make progress on previously agreed action plans (Bali) and agreements (Cancun). The 17th Conference of the Parties (COP 17) to the UNFCCC and the 7th Session of the COP serving as the Meeting of the Parties (MOP 7) to the Kyoto Protocol were, however, especially from an African viewpoint, only partly successful. Not only were China and the United States reluctant to sign binding targets, each waiting for the other to move first, but this reluctance by the two biggest polluters had repercussions for the negotiations as a whole, namely that Canada, Japan and Russia refused to enter into a second commitment period for the Kyoto Protocol owing to the lack of legal restriction on the world's largest polluters. The sense of dampened success applies even more to the outcomes of the 18th COP 18 to the UNFCCC and the 8th Session of the COP serving as the Meeting of the Parties (MOP 8) to the Kyoto Protocol held in Doha, Qatar, in 2012, where a number of decisions were adopted (The Doha Climate Gateway).[8] A second commitment period under the Kyoto Protocol has been launched, with the end date being 2020.[9] It has been agreed to work towards a universal climate change agreement covering all countries from 2020. Such agreement is to be adopted by 2015. Countries have furthermore agreed on ways and means to deliver scaled-up climate finance and technology to developing countries, and COP18 has also taken note of the first annual report of the Board of the Green Climate Fund to the Conference of the Parties and endorsed the consensus decision of the Board of the Green Climate Fund to select Songdo, Incheon, in the Republic of Korea as the host city of the Green Climate Fund, on the basis of an open and transparent process.[10] Further key elements of the outcome include an agreement to consider loss and damage in developing countries that are particularly vulnerable to the adverse effects of climate change.

---

6 AMCEN/African Ministerial Conference on the Environment, 2011, *Addressing Climate Change Challenges in Africa: A Practical Guide towards Sustainable Development*, retrieved on 15 January 2013 from <www.unep.org/roa/amcen/docs/.../guidebook_CLimateChange.pdf>.

7 *See also* <www.dieburger.com/Suid-Afrika/Nuus/Krisis-raak-net-erger-20111129> (last accessed 23 December 2011).

8 All decisions adopted by COP 18 and CMP 8 can be accessed at <http://unfccc.int/meetings/doha_nov_2012/meeting/6815.php#decisions> (last accessed 16 January 2013).

9 However, some previously participating countries in the Kyoto Protocol have not joined the second period, namely Russia, Canada, New Zealand and Japan.

10 *See* UNFCCC/United Nations Framework Convention on Climate Change, 2012, Report of the Green Climate Fund to the Conference of the Parties and Guidance to the Green Climate Fund, available at <http://unfccc.int/resource/docs/2012/cop18/eng/05.pdf> (last accessed 18 February 2013).

"At the Doha conference, Germany, the UK, France, Denmark, Sweden and the EU Commission announced financial pledges for the period up to 2015 totalling approximately $6 billion. Most developed countries did, however, not make pledges! African countries thus left Doha with little more than they already had."[11] This means that for Africa climate change continues to prompt significant challenges in future,[12] and it is therefore noteworthy that the AU and a number of African countries have embarked on new policy pathways to accommodate climate change more effectively.

4.2 The African Union

The historical foundations of the AU originated in the Union of African States, an early confederation that was established in the 1960s. The Organization of African Unity (OAU) was established on 25 May 1963. On 9 September 1999, the heads of state and governments of the OAU issued the Sirte Declaration,[13] calling for the establishment of an AU. The Declaration was followed by summits in Lomé in 2000, when the Constitutive Act of the African Union was adopted, and in Lusaka in 2001, when the Plan for the Implementation of the African Union was adopted. During the same period, the initiative for the establishment of the New Partnership for Africa's Development (NEPAD) was also instituted.[14] The AU was launched in Durban on 9 July 2002 by the then South African president, Thabo Mbeki,[15] at the First Session of the Assembly of the AU. The Union's administrative centre is in Addis Ababa, Ethiopia, and its working languages are Arabic, English, French, Portuguese and Swahili. The AU has 54[16] Member States, Morocco being the only African state that is not a member. Geographically, the AU covers an area of 29,757,900 km^2, and for 2010, the United Nations Population Division estimated a population total of 990,283,000.[17]

Article 3 of the Constitutive Act of the African Union contains the objectives of the AU, which include the promotion of sustainable development, international cooperation and continental integration, as well as the promotion of scientific and technological research to advance the continent's development. In the Protocol relating to the Establishment of

11 *See* <www.un.org/africarenewal/web-features/what-does-'doha-climate-gateway'-mean-africa> (last accessed 19 January 2013).

12 O.C. Ruppel & S. van Wyk, 'The Effects of Climate Change on Human Security and Human Rights in Africa', *China Monitor*, 2011, available at <www.ccs.org.za/wp-content/uploads/2011/10/China_Monitor_OCT_2011_final.pdf> (last accessed 23 December 2012).

13 Named after Sirte in Libya.

14 O.C. Ruppel, 'International Environmental Law from a Namibian Perspective', *in* O.C. Ruppel & K. Ruppel-Schlichting (Eds.), *Environmental Law and Policy in Namibia*, Windhoek, Orumbonde Press, 2011a, p. 43.

15 Thabo Mbeki was also the AU's first president.

16 Including South Sudan, which ratified the Constitutive Act of the African Union on 15 August 2011; *see* <www.au.int/en/sites/default/files/Constitutive_Act_0.pdf> (last accessed 28 March 2012).

17 Africa's entire population was estimated to be 1,022,234,000. As Morocco is not a member of the AU, its population – estimated at 31,951,000 – has to be deducted. Data collected from <http://esa.un.org/unpd/wpp/unpp/panel_population.htm> (last accessed 28 March 2012).

the Peace and Security Council (PSC) of the AU, Member States committed themselves to various guiding principles (Art. 4), including "early responses to contain crisis situations," and the recognition of the "interdependence between socio-economic development and the security of peoples and States". Moreover, in Article 6 of the Constitutive Act, the functions of the PSC are outlined as, among other things, the promotion of peace, security and stability in Africa; early warning and preventive diplomacy; peacemaking; humanitarian action; and disaster management. All the aforementioned provisions give a clear mandate to address climate change as a matter of priority within the AU.

4.3 The New Partnership for Africa's Development

NEPAD was adopted in 2001 in Lusaka, Zambia, by heads of state and government of the OAU in 2001. NEPAD was ratified by the AU in 2002. Partnership and cooperation between Africa and the developed world are envisaged by this development initiative aimed at the economic and social revival of Africa. NEPAD is[18]

> [. . .] a pledge by African leaders, based on a common vision and a firm and shared conviction, that they have a pressing duty to eradicate poverty and to place their countries, both individually and collectively, on a path of sustainable growth and development, and at the same time to participate actively in the world economy and body politic. The Programme is anchored on the determination of Africans to extricate themselves and the continent from the malaise of underdevelopment and exclusion in a globalising world.

NEPAD has emphasized Africa's important role in respect of the critical issue of environmental protection. Of the six main thematic areas on which NEPAD focuses, two are of particular relevance here[19]

- Agriculture and food security, and
- Climate change and natural resource management.

NEPAD's Climate Change and Natural Resource Management Programme focuses on three key areas: environment, water and energy. The Programme aims to assist countries in integrating climate change responses with their national development processes. The Programme also aims to strengthen skills in adaptation, mitigation, technology and finance in order to combat environmental change. To achieve these aims, an Action Plan for the Environment Initiative was released in 2003, paving the way for the first decade of the 21st century.[20]

---

18 *See* NEPAD's founding document, available at <www.dfa.gov.za/au.nepad/nepad.pdf> (last accessed 12 February 2012).

19 Further thematic areas within the NEPAD framework are Regional Integration and Infrastructure, Human Development, Economic and Corporate Governance, and Crosscutting Issues (Gender and Capacity Management).

20 Available at <www.nepad.org/system/files/Environment%20Action%20Plan.pdf> (last accessed 12 February 2012).

4.4 THE AU'S JUDICIAL SYSTEM AND CONSIDERATION OF ENVIRONMENTAL RIGHTS

In 1998, the African Court on Human and Peoples' Rights (AfCHPR) was established by the Protocol to the African Charter on Human and Peoples' Rights on the Establishment of an African Court on Human and Peoples' Rights, which came into force in 2004. The AfCHPR is situated in Arusha, in the United Republic of Tanzania, and has received cases since June 2008. In 2003, the African Court of Justice – as the ultimate organ of jurisdiction in the AU – was established by the Protocol of the Court of Justice of the African Union, which came into force in February 2009. However, the Protocol on the Statute of the African Court of Justice and Human Rights adopted in 2008 during the AU Summit of Heads of State and Government in Sharm El Sheikh in the Arab Republic of Egypt provides that the 1998 and 2003 Protocols be replaced, and that the AfCHPR and the AU's Court of Justice be merged into a single court to become what is now known as the African Court of Justice and Human Rights. However, the 2008 Protocol on the merger of the courts has so far been ratified only by five[21] Member States, but ratification by a minimum of fifteen is required for the Protocol to come into force. Once operational, the merged court will have two sections: a General Affairs Section and a Human Rights Section, both composed of eight Judges. The court will have jurisdiction over all disputes and applications referred to it, which, among other things, relate to the interpretation and application of the AU Constitutive Act or the interpretation, application and validity of Union Treaties, as well as human rights violations.

The African Commission on Human and Peoples' Rights (hereafter *African Commission*) is a quasi-judicial body established by the 1981 African (Banjul) Charter on Human and Peoples' Rights (hereafter *African Charter*) and is responsible for monitoring compliance with the African Charter. The African Charter is a human rights treaty that already proclaims environmental rights in broadly qualitative terms. It protects the right of peoples both to the "best attainable state of physical and mental health" (Art. 16) and to a "general satisfactory environment favourable to their development" (Art. 24). Article 24 of the African Charter further establishes a binding human rights based approach to environmental protection, linking the right to environment to the right to development.[22]

In the *Endorois* case,[23] the African Commission concluded that several articles of the African Charter had been violated in the course of the dispossession of the land of Kenya's indigenous Endorois through the creation of the Lake Hannington Game Reserve in 1973, and a subsequent regazetting of the Lake Bogoria Game Reserve in 1978 by the Kenyan

21 As of 1 March 2013, the Protocol was ratified by Benin, Burkina Faso, Congo, Libya and Mali. *See* <www.au.int/en/sites/default/files/Protocol%20on%20Statute%20of%20the%20African%20Court%20of%20Justice%20and%20HR.pdf> (last accessed 7 April 2013).

22 M. Van der Linde & L. Louw, 'Considering the Interpretation and Implementation of Article 24 of the African Charter on Human and Peoples' Rights in Light of the SERAC Communication', *African Human Rights Law Journal*, Vol. 3, No. 1, 2003, pp. 167-187.

23 Communication 276/03, *Centre for Minority Rights Development (Kenya) and Minority Rights Group International on behalf of Endorois Welfare Council v. Kenya*, available at <www.achpr.org/english/Decison_Communication/Kenya/Comm.%20276-03.pdf> (last accessed 12 February 2012).

Government. Among the Endorois' rights found to be violated were their right to culture (Art. 17(1) and (2)) and their right to the free disposition of natural resources (Art. 21), as they were unable to access the vital resources in the Lake Bogoria region after their eviction from the game reserve. Moreover, the African Commission held that the Endorois' right to development (Art. 22) had been violated, as the respondent state had failed to adequately involve the Endorois in the development process.[24] The decision of the African Commission in the *Endorois* case[25] was influenced by provisions of Convention No. 169 of the International Labour Organization on Indigenous and Tribal Peoples in Independent Countries.[26] Among other things, the Convention[27]

- provides criteria for describing the peoples it aims to protect
- makes provision regarding the principle of non-discrimination
- calls for special measures to be adopted to safeguard the persons, institutions, property, labour, cultures and environment of indigenous and tribal peoples
- recognizes cultural and other specificities of indigenous and tribal peoples, and
- requires that, on all issues that affect them, indigenous and tribal peoples are consulted, and that these peoples are able to engage in free, prior and informed participation in policy and development processes.

In the *Ogoni* case, the African Commission held, among other things, that Article 24 of the African Charter imposed an obligation on the state to take reasonable measures to "prevent pollution and ecological degradation, to promote conservation, and to secure ecologically sustainable development and use of natural resources."[28]

The *Ogoni* case led to a landmark decision with regard to the effective protection of economic, social and cultural rights in Africa, particularly the protection of the right of peoples to a satisfactory environment.[29] Article 24 of the African Charter should be

---

24 The Commission's recommendation was to recognize the Endorois' rights of ownership and to restore their ancestral land, to ensure that the Endorois community had unrestricted access to Lake Bogoria and surrounding sites for religious and cultural rites and for grazing their cattle, to pay adequate compensation to the community for all the loss they had suffered, to pay royalties to the Endorois from existing economic activities and ensure that they benefited from employment possibilities within the Reserve, to grant registration to the Endorois Welfare Committee and to engage in dialogue with the complainants for the effective implementation of these aforementioned recommendations and report on their implementation.

25 Communication 276/03, *Centre for Minority Rights Development (Kenya) and Minority Rights Group International on behalf of Endorois Welfare Council v. Kenya*, available at <www.achpr.org/english/Decison_Communication/Kenya/Comm.%20276-03.pdf> (last accessed 28 March 2012).

26 The Convention came into force on 5 September 1991 and is available at <www.ilo.org/ilolex/cgi-lex/convde.pl?C169> (last accessed 12 February 2012).

27 It should be noted that of the 22 states that have ratified ILO Convention No. 169, as of February 2012, only one – the Central African Republic – is from the African continent.

28 Communication 155/96, *The Social and Economic Rights Action Center (SERAC) & the Center for Economic and Social Rights (CESR) v. Nigeria*, available at <www.achpr.org/english/_info/decision_article_24.html> (last accessed 28 March 2012).

29 O.C. Ruppel, 'Human Rights and the Environment', *in* O.C. Ruppel & K. Ruppel-Schlichting (Eds.), *Environmental Law and Policy in Namibia*, Windhoek, Orumbonde Press, 2011b, pp. 219-237.

viewed together with the Bamako Convention and the first OAU treaty on the environment, namely the Convention on the Conservation of Nature and Natural Resources, which predates the African Charter. The Revised African Convention on the Conservation of Nature and Natural Resources was adopted by the Second Ordinary Session of the AU Assembly of Heads of State and Government in Maputo, Mozambique, in July 2003. However, the latter Convention has not yet come into force.

The recognition in the African Charter of a right to a satisfactory environment and the progressive jurisprudence by the African Commission take up the issue of environmental protection from a human rights perspective and underline the linkage between climate change and human rights in a modern, holistic approach to one of today's burning issues.[30] The impacts of climate change on human rights have been explicitly recognized by the African Commission: in its AU Resolution 153, the African Commission calls on the Assembly of Heads of State and Government to ensure that[31]

> human rights standards safeguards, such as the principle of free, prior and informed consent, be included into any adopted legal text on climate change as preventive measures against forced relocation, unfair dispossession of properties, loss of livelihoods and similar human rights violations;

and "to take all necessary measures to ensure that the African Commission on Human and Peoples' Rights is included in the AU's negotiating team on climate change." In the same communication, the African Commission resolves to carry out a study on the impact of climate change on human rights in Africa.[32]

4.5 AU Climate Change Policy and Related Developments

Many African governments have made progress in addressing climate change and related issues. The AU itself has succeeded in presenting a more cohesive African position on climate change. Although gaps may remain, it was clearly reflected during COP 17 in Durban that Africa maintained a common position in spite of pressure from developed countries. In fact, the African Group "spoke with one voice," according to Seyni Nafo, spokesperson for the African Group and lead negotiator on Mitigation.[33] It has since become apparent,

30 O.C. Ruppel, 'Environmental Rights and Justice in Namibia', *in* A. Bösl, N. Horn & A. du Pisani (Eds.), *Constitutional Democracy in Namibia: A Critical Analysis after Two Decades*, Windhoek, Macmillan Education Namibia, 2010a, pp. 323-360.

31 AfCHPR/Res. 153 (XLV09).

32 *See* <www.achpr.org/english/resolutions/resolution153_en.htm> (last accessed 14 February 2012).

33 *See* <www.dailytrust.com.ng/index.php?option=com_content&view=article&id=149432:africa-maintains-common-position-in-durban&catid=10:environment&Itemid=11> (last accessed 20 February 2012).

however, that divergent priorities among African countries threaten the potential of the AU to influence international climate politics.[34]
At its 46th Ordinary Session, held in Banjul, The Gambia, from 11 to 25 November 2009, the African Commission urged[35]

> [...] the Assembly of Heads of State and Government of the African Union to ensure that human rights standards safeguards, such as the principle of free, prior and informed consent, be included into any adopted legal text on climate change as preventive measures against forced relocation, unfair dispossession of properties, loss of livelihoods and similar human rights violations.

4.5.1 *Climate Change Induced Displacement*

One legal instrument that explicitly deals with the potential impacts of climate change is the African Union Convention for the Protection and Assistance of Internally Displaced Persons in Africa (hereafter *Kampala Convention*),[36] which was adopted in Kampala on 23 October 2009. It is the first regional legal instrument in the world to contain legal obligations for states, with regard to the protection and assistance of internally displaced persons (IDPs).[37] So far, the Kampala Convention has 39 signatories, and 17 countries[38] have ratified it. The Convention entered into force on 6 December 2012.[39] Article 1(k) of the Kampala Convention defines IDPs as follows:

> [P]ersons or groups of persons who have been forced or obliged to flee or to leave their homes or places of habitual residence, in particular as a result of or in order to avoid the effects of armed conflict, situations of generalized violence, violations of human rights or natural or human-made disasters, and who have not crossed an internationally recognized State border.

The Kampala Convention explicitly recognizes climate change as one of the possible reasons for internal displacement: Article 5 states that "States Parties shall take measures to

---

34 J-C. Hoste, 'Where Was United Africa in the Climate Change Negotiations?', *Africa Policy Brief* 2, Brussels, Egmont Royal Institute for Internatonal Relations, available at <http://www.edc2020.eu/fileadmin/Textdateien/post_COP_15_briefing/Jean_Christophe_Hoste_-_Where_was_united_Arfrica_in_the_climate_change_negotiations_-_EDC_2020.pdf>) (last accessed 13 January 2013), 2010.

35 *See* <www.achpr.org/english/resolutions/resolution153_en.htm> (last accessed 14 February 2012).

36 Text available at <www.au.int/en/sites/default/files/AFRICAN_UNION_CONVENTION_FOR_THE_PROTECTION_AND_ASSISTANCE_OF_INTERNALLY_DISPLACED_PERSONS_IN_AFRICA_(KAMPALA_CONVENTION).pdf>(last accessed 30 January 2012).

37 *See* W. Kidane, 'Managing Forced Displacement by Law in Africa: The Role of the New African Union IDPs Convention', 44 *Vanderbilt Journal of Transnational Law* 1, 2011, pp. 1-85.

38 As of 7 April 2013, the following seventeen Member States had ratified the Convention: Benin, Burkina Faso, Central African Republic, Chad, Gabon, Gambia, Guinea-Bissau, Lesotho, Mali, Nigeria, Niger, Rwanda, Sierra Leone, Swaziland, Togo, Uganda and Zambia, *see* <www.au.int/en/sites/default/files/Convention%20on%20IDPs%20-%20displaced....pdf> (last accessed 7 April 2013).

39 Ratification of fifteen Member States was required for the Convention to come into force.

protect and assist persons who have been internally displaced due to natural or human-made disasters, including climate change."

However, the Kampala Convention applies to all situations of internal displacement regardless of its causes (Art. 15), which makes sense: drivers of migration in general are not mono-causal, but are influenced by multiple factors.

Several obligations are imposed on states parties by the Kampala Convention. For example, the Convention addresses the need to prevent displacement from happening, *e.g.* by establishing early warning systems and adopting disaster preparedness and management measures to prevent displacement caused by natural disaster. The Convention also requires States Parties to, among other things, protect people against displacement resulting from conflict and violence, discriminatory policies, or human rights violations. Neither should displacement be used as a method of warfare, nor as collective punishment. Forced evacuations should only take place for reasons of health and safety, and, once people have been displaced, the Kampala Convention provides that they are to be protected and assisted (Article 4). According to Article 5, States Parties are obliged to assess the needs and vulnerabilities of displaced persons, as well as those of the host communities, and to provide adequate assistance, if need be with assistance from relevant local and international agencies. One objective of the Kampala Convention is to provide for durable solutions with respect to IDPs, who retain the right to make a free and informed choice on whether to return, integrate or relocate elsewhere in the country (Art. 11). Furthermore, States Parties are responsible for establishing an effective legal framework to provide just and fair compensation, and other forms of reparations for damage incurred as a result of displacement (Art. 12).

In light of the aforementioned, it is also worthwhile to note that various sub-regional organizations, *i.e.* Regional Economic Communities (RECs), have "at the insistence of the AU" established certain mechanisms.[40] At the seventh ordinary session of the AU's Assembly of Heads of State and Government in Banjul, The Gambia, in July 2006, the AU officially recognized eight such communities.[41] Alphabetically listed, these are as follows:[42]

- The Arab Maghreb Union (AMU);
- The Community of Sahel-Saharan States (CEN-SAD);
- The Common Market for Eastern and Southern Africa (COMESA);
- The East African Community (EAC);
- The Economic Community of Central African States (ECCAS);
- The Economic Community of West African States (ECOWAS);

---

40 *See* O.C. Eze, 'The Conventions Governing the Specific Aspects of Refugee Problems in Africa', *in* A.A. Yusuf & O. Fatash (Eds.), *The African Union: Legal and Institutional Framework*, Leiden; Boston, Martinus Nijhoff Publishers, 2012, p. 516.

41 *See* the decision relating to the recognition of RECs, Assembly/AU/Dec.112 (VII) Doc. EX.CL/278 (IX), text in French, available at <www.africa-union.org/Official_documents/Assemblee%20fr/ASS06b.pdf> (last accessed 22 December 2012).

42 O.C. Ruppel, 'Regional Economic Communities and Human Rights in East and Southern Africa', *in* A. Bösl & J. Diescho (Eds.), *Human Rights in Africa*, Windhoek, Macmillan Publishers, 2009a, p. 276.

Table 1 State Members of RECs Officially Recognized by the AU[43]

Amu	Cen-sad	Comesa	Eac	Eccas	Ecowas	Igad	Sadc
Algeria	Benin	Burundi	Burundi	Angola	Benin	Djibouti	Angola
Libya	Burkina Faso	Comoros	Kenya	Burundi	Burkina Faso	Ethiopia	Botswana
Mauritania	Central African Republic	DRC	Rwanda	Cameroon	Cape Verde	Kenya	DRC
Morocco	Chad	Djibouti	Tanzania	Central African Republic	Cote d'Ivoire	Somalia	Lesotho
Tunisia	Comoros	Egypt	Uganda	Chad	Gambia	Sudan	Madagascar
	Cote d'Ivoire	Eritrea		Congo	Ghana	Uganda	Malawi
	Djibouti	Ethiopia		DRC	Guinea		Mauritius
	Egypt	Kenya		Gabon	Guinea-Bissau		Mozambique
	Eritrea	Libya		Guinea	Liberia		Namibia
	Gambia	Madagascar		São Tomé and Príncipe	Mali		Seychelles
	Ghana	Malawi			Niger		South Africa
	Guinea-Bissau	Mauritius			Nigeria		Swaziland
	Bissau						
	Kenya	Rwanda			Senegal		Tanzania
	Liberia	Seychelles			Sierra Leone		Zambia

Table 1 State Members of RECs Officially Recognized by the AU

Amu	Cen-sad	Comesa	Eac	Eccas	Ecowas	Igad	Sadc
	Libya	Sudan			Togo		Zimbabwe
	Mali	Swaziland					
	Mauritania	Uganda					
	Morocco	Zambia					
	Niger	Zimbabwe					
	Nigeria						
	São Tomé and Príncipe						
	Senegal						
	Sierra Leone						
	Somalia						
	Sudan						
	Togo						
	Tunisia						

43 *Ibid.*, p. 278.

- The Intergovernmental Authority on Development (IGAD), and
- The Southern African Development Community (SADC).

All AU Member States are affiliated to one or more of these RECs, as tabulated below:

At first glance, it appears that the promotion and protection of displaced persons is not within the RECs' focal range. However, it will still have to be seen how effective these mechanisms actually are. Regional integration, which is the primary engine of RECs, certainly has the potential to provide[44]

> [. . .] an opportunity to enhance political stability by establishing regional **organisations which play an increasing role in defusing conflicts within** and between countries and in promoting human rights. In terms of climate change related matters, such organisations are of the utmost relevance, especially when it comes to climate change related disaster management and environmentally induced migration.

4.5.2 Climate Change and Vulnerable Groups

The African Commission in 2009 urged[45]

> [. . .] the Assembly of Heads of State and Government to ensure that special measure of protection for vulnerable groups such as children, women, the elderly, indigenous communities and victims of natural disasters and conflicts are included in any international agreement or instruments on climate change.

This call is very much in line with Article 3 of the International Covenant on Economic, Social and Cultural Rights,[46] the Convention on the Elimination of All Forms of Discrimination against Women, and the latter's Optional Protocol.[47] The vulnerability of women to climate change and natural disasters is more severe than for other groups for a number of reasons. Women are usually at higher risk of being placed in unsafe, overcrowded shelters owing to a lack of assets such as savings, property or land. In the context of droughts, floods and other disasters that require mobility, cultural constraints on women's movements may

44 *See* O.C. Ruppel & K. Ruppel-Schlichting, 'Climate Change and Human Security: Relevant for Regional Integration in SADC?', *in* T. Hartzenberg, G. Erasmus & A. du Pisani (Eds.), *Monitoring Regional Integration in Southern Africa, 2011 Yearbook for Regional Integration*, Stellenbosch, Trade Law Centre for Southern Africa (TRALAC), 2012, p. 41.

45 *See* <www.achpr.org/english/resolutions/resolution153_en.htm> (last accessed 14 February 2012).

46 Art. 3 encourages States Parties to ensure the equal right of men and women to the enjoyment of all economic, social and cultural rights as set forth by the Covenant.

47 *See* Ruppel, 2008; Ruppel, 2010b.

hinder their timely escape, access to shelter, or access to health care. These effects are exacerbated when women avoid using shelters out of fear of domestic and sexual violence, and become even less mobile as primary family caregivers. Poor women and those in countries with greater gender inequality appear to be most at risk.[48]

The impact of climate change on the realization of child rights[49] as a whole obviously also includes multiple effects on basic rights such as water, food and health, especially in African countries that are vulnerable to temperature and precipitation change. Children are vulnerable to climate change, and policy makers need to adhere to the Convention on the Rights of the Child that requires national governments to ensure that children's specific needs are given due consideration in adaptation and mitigation policy.[50]

It is worth noting that Cancun Decision 1/CP.16 also recognizes indigenous peoples as a vulnerable group, alongside women, children and other vulnerable groups. In Africa, indigenous peoples are vulnerable to the actual and potential detrimental impacts of climate change. In line with the Declaration on the Rights of Indigenous Peoples, which was adopted by the United Nations General Assembly in 2007,[51] indigenous peoples should be free from discrimination of any kind – including that in the context of climate change.

4.5.3 Climate Change, Peace and Collective Security

The AU has a clear mandate regarding the maintenance of peace and security in Africa. Yet, Africa remains a continent blighted by conflict, where "millions of human beings remain at the mercy of civil wars, insurgencies, state repression and state collapse."[52] But these are not the only challenges. Climate change overlies these ravages, intensifying existing problems as a magnifying glass would. Although climate change is, in the first instance, an environmental issue, it is also political, with "far-reaching economic, societal and political ramifications,"[53] which cannot be neglected.

In fact, framing climate change as more of a security issue in Africa could serve to enhance existing policy response mechanisms. In 2011, the United Nations Security Council

48 UN/United Nations Womenwatch, *Fact Sheet: Women and Climate Change,* 2009, available at <www.un.org/womenwatch/feature/climate_change/downloads/Women_and_Climate_Change_Factsheet.pdf> (last accessed 25 June 2011).

49 *See* O.C. Ruppel, 'The Protection of Children's Rights under International Law from a Namibian Perspective', *in* O.C. Ruppel (Ed.), *Children's Rights in Namibia,* Windhoek, Macmillan Publishers Namibia, 2009b.

50 UNICEF/United Nations Children's Fund, *A Brighter Tomorrow: Climate Change, Child Rights and Intergenerational Justice,* available at <www.unicef.org.uk/Documents/Publications/intergenerationaljustice.pdf> (last accessed 25 June 2011).

51 Resolution 61/295.

52 Report of the International Commission on Intervention and State Sovereignty, 2011.

53 *See* S.V. Scott, 'The Securitization of Climate Change in World Politics: How Close Have We Come and Would Full Securitization Enhance the Efficacy of Global Climate Change Policy?', *Review of European Community and International Environmental Law,* Vol. 21, No. 3, 2012, p. 220.

expressed concern that the possible adverse effects of climate change could, in the long run, intensify certain existing threats to international peace and security and that the loss of territory in some states could have possible security implications.[54]

The Constitutive Act of the African Union, which was adopted in July 2000,[55] stipulates the Union's right of intervention in a Member State in cases of "grave circumstances," *i.e.* crimes against humanity, war crimes and genocide. This raises the question of how such grave circumstances relate to climate change. Notwithstanding the controversy around this question, Achim Steiner, when addressing the UN Security Council in the 2011 Debate on Climate Change and Security, clearly stated the following:[56]

> **Climate change as a threat multiplier**: The scale and pace of climate change acts as a multiplier which could result in simultaneous and unprecedented impacts on where we can settle, grow food, maintain our built-up infrastructure, or rely on functioning ecosystems. Managing the potential disruption, displacement and adaptation to phenomena such as sea-level rise or extreme weather events, represents a profound challenge to sustainable development at the local, national and international level – both in economic and geopolitical terms. [. . .] In 2010, over 90 per cent of disaster displacement within countries was caused by climate-related hazards, primarily floods and storms. Climate scenarios expect such weather events to increase and/or intensify as a result of accelerating climate change. [. . .] Competition over scarce water and land, exacerbated by regional changes in climate, are already a key factor in local-level conflicts in Darfur, the Central African Republic, northern Kenya, and Chad, for example-when livelihoods are threatened by declining natural resources, people either innovate, flee or can be brought into conflict.

The aforementioned statements clearly pinpoint climate change as a potential source of conflict and a potential threat to national and international peace and human security. One cornerstone of the United Nations Charter paradigm is the notion of collective security, which is perhaps the first and most obvious manifestation of the principle of solidarity in the post-World War II era.[57] In fact, it forms the political and legal foundation for the collective

54 Security Council Meeting SC/10332 document available at <www.un.org/News/Press/docs/2011/sc10332.doc.htm> (last accessed 18 February 2013).

55 Constitutive Act of the African Union, 11 July 2000.

56 Address by UN Under-Secretary General and UNEP Executive Director Achim Steiner at the UN Security Council Debate on the impact of climate change on maintaining international peace and security, 20 July 2011, available at <www.unep.org/Documents.Multilingual/Default.Print.asp?DocumentID=2646&ArticleID=8817&l=en>(last accessed 18 February 2013).

57 *See* A.D. Koroma, 'Solidarity: Evidence of an Emerging International Legal Principle', *in* H.P. Hestermeyer *et al.* (Eds.), *Coexistence, Cooperation and Solidarity: Liber Amicorum Rüdiger Wolfrum*, Leiden, Martinus Nijhoff Publishers, 2012.

security system established by the UN Charter. A still controversial manifestation of the notion of solidarity in international law is the emerging doctrine of the responsibility to protect. This concept was developed by the International Commission on Intervention and State Sovereignty in September 2000, after UN Secretary General Kofi Annan emphasized the grave failure of the international community to handle gross and systematic violations of human rights such as those perpetrated in Rwanda and others.[58] The aforementioned concept has gained growing attention in the context of the notion of global solidarity and collective security as it aims at addressing legal and political dilemmas for intervention to stop or prevent human suffering and crimes against humanity.[59]

The PSC of the African Union in its recent report[60] documented its activities and the state of peace and security in Africa pursuant to Article 7(q) of the Protocol Relating to the Establishment of the PSC of the AU. It provides an overview of the state of peace and security on the continent from July 2012 to January 2013. During this reporting period, the PSC considered the crisis and conflict situations in the following states: Central African Republic, Sudan, Democratic Republic of Congo, Guinea Bissau, Mali and Somalia, as well as the situation between Sudan and South Sudan. During the reporting period, the PSC considered several thematic issues relating to the promotion of peace, security and stability in Africa, such as the following:[61]

- Capacity building for effective response to humanitarian assistance and disasters in Africa
- Challenges related to peace and security in Africa from the perspective of delivering humanitarian assistance in situations of crisis and conflict
- The need to mainstream gender in all development, peace and security efforts on the continent, and
- The importance of international and transitional justice in the promotion of peace and security in Africa.

The PSC in its report furthermore[62]

- reiterated the commitment of the AU to the fight against impunity, and stressed the importance of international and transitional justice in the promotion of peace and security in Africa, and the need, in the context of the search for solutions to crises and conflicts and in view of the fragility of the peace and reconciliation processes on the continent, to ensure that they are mutually reinforcing; and

58 Report of the Secretary-General on the Work of the Organization, document A/54/1, p. 48, 31 August 1999.

59 *See* Koroma, 2012.

60 Assembly of the African Union, 20th Ordinary Session, 27-28 January 2013, Addis Ababa, Ethiopia, Assembly/AU/3(XX), available at <www.peaceau.org/uploads/assemblyau-3-xx-e.pdf> (last accessed 5 April 2013).

61 *Ibid.*

62 *Ibid.*

- underscored the fact that the primary responsibility for the protection of human rights rests with Member States;
- emphasized the need for a close working relationship with the African Commission on Human and Peoples' Rights, as well as with the African Court on Human and Peoples' Rights.

Under Article 52 of the UN Charter, regional organizations may undertake actions aimed at the maintenance of international peace and security. Article 53(I) of the UN Charter specifically provides that such regional organizations may undertake enforcement measures, provided that they have the authorization of the UN Security Council. As mentioned earlier, Article 4(h) of the Constitutive Act of the African Union provides for such intervention in the event of grave circumstances, and it also gives the Assembly of the African Union the authority to decide over it. Moreover, Article 13(3)(c) of the African Union Peace and Security Protocol provides a mandate to the African Standby Force to intervene in the activities of a Member State, when grave circumstances demand such intervention, in accordance with Article 4(h) of the Constitutive Act of the African Union.[63]

Most obviously, the responsibility to protect the concept strongly revolves around the dilemma between state sovereignty and intervention for humanity. In light of this, current discussions focus on the duty of the international community and the territorial state in cases of natural disasters, which raises the question whether the doctrine of the responsibility to protect can actually be extended to the international law relating to disaster relief and in particular to cases of grave circumstances, such as severe human suffering during times of natural disasters.

Unfortunately, international law and politics thus far still seem to attach great significance to whether human suffering is the result of a natural disaster or of an (international) armed conflict.[64] However, when responding to the question whether the doctrine of the responsibility to protect should be extended to international law relating to disaster relief, one should argue in the interest of the African people. Once again, consider the words of Achim Steiner:[65]

> There is no reason why the international community cannot avoid escalating conflicts, tensions and insecurity related to a changing climate if a deliberate, focused and collective response can be catalyzed that tackles the root causes,

63 *See* T. Kabau, 'The Responsibility to Protect and the Role of Regional Organizations: An Appraisal of the African Union's Interventions', *Goettingen Journal of International Law*, Vol. 4, 2012.

64 *See* P. Thielbörger & M. Liburd, *Haiti Two Years after the Earthquake – A Case for the Responsibility to Protect?*, Institute for International Law of Peace and Armed Conflict, BOFAXE No. 399E, 2012.

65 Address by UN Under-Secretary General and UNEP Executive Director Achim Steiner at the UN Security Council Debate on the impact of climate change on maintaining international peace and security, 20 July 2011, available at <www.unep.org/Documents.Multilingual/Default.Print.asp?DocumentID=2646&ArticleID=8817&l=en> (last accessed 19 February 2013).

scale, potential volatility and velocity of the challenges emerging. In bringing forward a response that enhances global security and cooperation on the climate challenge, the world can perhaps also better manage risk from numerous other challenges and in doing so diminish tensions between nations and lay the foundations and possibilities of a more sustainable and equitable peace.

4.6 The African Ministerial Conference on the Environment

The African Ministerial Conference on the Environment (AMCEN), which has so far played a prominent role in the African response to climate change,[66] has a strong regional and sub-regional focus. Thus, AMCEN builds on the potential for RECs to integrate adaptation measures into regional policies and socio-economic development.[67] AMCEN is a permanent forum, where African ministers of the environment discuss mainly matters of relevance to the environment on the continent. The forum was established in Egypt in 1985, the year in which the Cairo Programme for African Cooperation has been adopted. AMCEN has convened every second year since then. In 2010, at its Thirteenth Session, AMCEN adopted the Bamako Declaration on the Environment for Sustainable Development as its contribution towards providing political guidance and leadership on environmental management to Africa. AMCEN was established to

- provide advocacy for environmental protection in Africa
- ensure that basic human needs are met adequately and in a sustainable manner
- ensure that social and economic development is realized at all levels, and
- ensure that agricultural activities and practices meet the food security needs of the region.

The adequate response to these challenges needs to be aligned with national and regional strategies for development, poverty alleviation, economic growth, and the enhancement of human well-being, while increasing resilience to the physical impacts of climate change. Several bodies have identified opportunities and challenges in the transition to a green economy, with links to the achievement of the United Nations Millennium Development Goals, climate change and sustainable development. These bodies include the Meeting of African Heads of State and Government at the Seventeenth Session of the AU Summit held in Malabo, Equatorial Guinea, in July 2011; AMCEN's Fourth Special Session held in Bamako, Mali, in September 2011; and, most recently, the Seventh Session

66 2009 Nairobi Declaration on the African Process for Combating Climate Change, UNEP/12/9.

67 W. Scholtz, 'The Promotion of Regional Environmental Security and Africa's Common Position on Climate Change', *African Human Rights Law Journal*, Vol. 10, 2010.

of the Committee on Food Security and Sustainable Development, as well as the Africa Regional Preparatory Conference on Sustainable Development (Rio+20) held in Addis Ababa, Ethiopia, in October 2011. In recognition of AMCEN's mandate, which includes guidance in respect of key issues related to multilateral environmental agreements, African governments asked AMCEN to facilitate the provision of information to countries that would assist them with translating available climate science and current international climate policies in their effort to move towards their practical implementation in the context of sustainable development. To this end, AMCEN prepared a guide book with information on climate change matters, including science, governance, technological, financial and capacity-building needs, as well as opportunities for effective action that would lead towards sustainable development.[68]

4.7 The African Climate Policy Centre

The African Climate Policy Centre (ACPC) was established in 2010 as an integral part of the Climate for Development in Africa Programme, which is a joint initiative of the African Union Commission, the United Nations Economic Commission for Africa and the African Development Bank.

The ACPC has been established to develop into a hub for a demand-led knowledge base on climate change in Africa to address the impact of climate change by assisting Member States to elevate climate change into mainstream deliberations in their development strategies and programmes. To this end, the ACPC hosted the inaugural Climate Change and Development in Africa (CCDA) Conference between 17 and 19 October 2011 at the United Nations Conference Centre in Addis Ababa, Ethiopia. The theme for the conference was 'Development First: Addressing Climate Change in Africa', which reflected the need for integrating development and climate policies, and emphasized the importance of African ownership of policy formulation and decision-making processes. The inaugural conference built directly on the African Development Forum VII and many other forums, initiatives, activities and outcomes of initiatives, including AMCEN, the Conference of African Heads of State and Government on Climate Change, the UNFCCC and related instruments, the UN Secretary General's High-level Advisory Group on Climate Change Financing, the Global Climate Observation System and its sub-regional climate programme and the Africa-EU Climate Change Partnership. The CCDA Conference helped to position the Climate for Development in Africa Programme within this ever-broadening knowledge and institutional terrain, and ascertained how best it could facilitate the interaction between policy, research and practice. The overall objective of the Conference was to establish a forum for dialogue, enhance awareness raising, and mobilize effective commitment and action by

68 *See* ACMEN, 2011.

bringing together policy makers, academicians and practising stakeholders, with the aim of effectively mainstreaming climate change concerns into development policies, strategies, programmes and practices in Africa. The CCDA also aimed at strengthening Africa's position and participation in international climate change negotiations with a view to ensuring the continent's concerns and priorities are adequately reflected in a post-2012 international climate change regime.[69]

All the aforementioned developments are laudable and should be seen as being in the overall interest of the AU. These developments, although not always clearly concerted, reflect that the AU and its subsidiary bodies have acknowledged that climate change and its impacts constitute a pressing policy priority.

4.8 NATIONAL CLIMATE CHANGE RELATED LEGAL DEVELOPMENTS

In light of the aforementioned AU policy pathways and related developments, accommodating national constitutional and legislative stipulations as well as climate change policy developments is critical because these are the linchpin between global, continental and domestic action. When countries enact policies responsive to climate change, it is more likely than not that investment will follow, which in turn can open opportunities for development and political space for international cooperation. Interaction around policy responses at the national and international level generates incremental, structural and transformational change. As there is no universal solution to environmental climate change, coordination, participation and cooperation are critical for jointly achieving internationally agreed upon goals and targets, while also addressing national capacity deficits.[70]

4.8.1 The Democratic Republic of Congo[71]

Noteworthy legal developments in environmental protection in the Democratic Republic of Congo (DRC) commenced with the adoption of a new constitution on 18 February 2006.[72] During the 45 years preceding its 2006 Constitution, the DRC introduced eight

69 AMCEN/African Ministerial Conference on the Environment, *Addressing Climate Change Challenges in Africa: A Practical Guide towards Sustainable Development*, 2012, www.unep.org/roa/amcen/docs/.../guidebook_CLimateChange.pdf (last accessed 15 January 2013).

70 *See* <www.unep.org/geo/pdfs/GEO5_SPM_English.pdf> (last accessed 19 January 2013).

71 Based on O.C. Ruppel & D. Bwiza, 'Country Report: Democratic Republic of Congo: Recent Developments in Environmental Protection', *IUCN E-Journal*, Vol. 4, 2013, pp. 138-147.

72 The adoption of a new constitution was a result of the overthrow of President Mobutu in 1996 by the opposition coalesced in the *Alliance des Forces Democratiques pour la Liberation du Congo-Zaire*. After a political impasse that lasted from 1997 to 2003, and the adoption of a government of national unity that ruled the country between 2003 and 2006, the DRC adopted a new constitution, which was followed by presidential elections.

previous constitutions.[73] The 2006 Constitution introduced clear environmental rights and obligations as follows:
- right to clean drinking water (Art. 48);
- right to a healthy environment and the duty to protect the environment (Art. 53);
- obligation of the state to protect the environment and to ensure health of populations (Art. 54), and
- obligatory control of domestic and international toxic waste resulting from economic activities (Art. 55).

Article 123 of the Constitution provides for laws to be made concerning, among others, the protection of the environment and tourism. To protect, among others, the environment, natural sites and landscapes, Article 203 allows for cooperative governance by central government and the provincial administrations.
The DRC alone accounts for one-fifth of Africa's total forest area, and as rainforests play a key role in the earth's carbon cycle,[74] it is seen to be particularly important to preserve these forests in the fight against climate change.[75] Besides many other International Environmental Agreements, the DRC is a party to the UNFCCC (since 1995) and the Kyoto Protocol (since 2005).
A National Climate Change Adaptation Plan (NAPA) was adopted in 2006, in order to develop a concrete priority action programme to guide the DRC's adaptation to climate change.[76] The NAPA is intended to provide the DRC with a framework to guide the coordination and implementation of climate change adaptation initiatives in the country, using a participative approach. The need to build synergies with other environmental programmes, such as the Biodiversity Action Plan, is emphasized. A national plan to combat desertification was finalized in 2012.[77] Various national environmental laws also relevant to the field of climate change have been enacted during the past years.[78]

---

73 Loi fondamentale du 19 juin 1960, Constitution du 1er août 1964, Constitution du 24 juin 1967, Loi constitutionnelle du 15 août 1974, Acte constitutionnel harmonisé du 2 avril 1993, Acte Constitutionnel de la transition du 9 avril 1994, Décret-loi constitutionnel du 27 mai 1997 et Constitution de la Transition du 3 avril 2003.

74 Food and Agricultural Organization of the United Nations, *State of the World's Forests 2011*, Rome, FAO 2011.

75 United Nations Environmental Programme (UNEP), *Africa Atlas of Our Changing Environment*, Nairobi, UNEP 2012, available at <http://unep.org/dewa/africa/africaAtlas/> (last accessed 28 December 2012).

76 Government of the Democratic Republic of Congo, *Programme d'Action National d'Adaptation au Changement Climatique de la République Démocratique du Congo*, 2012, available at <http://unfccc.int/resource/docs/napa/cod01.pdf> (last accessed 18 January 2012).

77 *See* <www.riddac.org/content/view/174/> (last accessed 15 November 2012).

78 That is, law no. 11/009 of 9 July 2011 on fundamental principles of environmental protection in the DRC on fundamental principles of environmental protection in the DRC; the Forest Code no. 011/2002 of 29 August 2002, which sets out the law applicable to the conservation, exploitation and development of forestry resources; the Mining Code law no. 007/2002 of 11 July 2002, which specifies the need for environmental impact assessments to be conducted for certain activities, such as prior feasibility studies of environmental, social and economic impact of projects, mitigation and rehabilitation plans, and environmental management plans for specific projects.

4.8.2 *Ethiopia*[79]

The Constitution of the Federal Democratic Republic of Ethiopia in Article 44 provides that

1. [a]ll persons have the right to a clean and healthy environment.
2. [a]ll persons who have been displaced or whose livelihoods have been adversely affected as a result of State programmes have the right to commensurate monetary or alternative means of compensation, including relocation with adequate State assistance.

Moreover, the Constitution stipulates the Environmental Objectives in Article 92 that

1. Government shall endeavour to ensure that all Ethiopians live in a clean and healthy environment.
2. The design and implementation of programmes and projects of development shall not damage or destroy the environment.
3. People have the right to full consultation and to the expression of views in the planning and implementation of environmental policies and projects that affect them directly.
4. Government and citizens shall have the duty to protect the environment.

Ethiopia is a non-Annex I member to the UNFCCC and the Kyoto Protocol, which it ratified in 2005. It has taken an active role in the recent climate negotiations and in the fight against deforestation in developing countries – mainly the poor and vulnerable ones. Ethiopia's 1995 Constitution includes the principle of environmental rights, including the right to a clean and healthy environment and the principle of government responsibility to ensure this right. In 2011, the Ethiopian Government finalized the Climate Resilient Green Economy[80] – the first of its kind in Africa. The strategy seeks to achieve ambitious economic development goals in a sustainable way by building a climate-resilient green economy.

Ethiopia's NAPA[81] was finalized in June 2007 by the Ministry of Water Resources and the Meteorological Service. The NAPA identifies high-priority adaptation projects, for example, promoting drought and crop insurance programmes; strengthening or enhancing drought and flood early warning systems; conserving and using wisely selected wetlands to promote the adaptation capacity of the rural community for climate shocks; enhancing the Community Based Carbon Sequestration Project in the Rift Valley System; establishing a national research and development (R&D) centre for climate change; strengthening

79 Based largely on GLOBE International, Climate Legislation Study: A Review of Climate Change Legislation in 33 Countries (3rd edn), *in* T. Townshend *et al.* (Eds.), 2103, available at <www.globeinternational.org/images/climate-study/3rd_GLOBE_Report.pdf>, pp. 134-141.

80 Federal Democratic Republic of Ethiopia, *Ethiopia's Climate-Resilient Green Economy Strategy*. Addis Ababa, Environmental Protection Agency 2011, available at <www.uncsd2012.org/content/documents/287CRGE%20Ethiopia%20Green%20Economy_Brochure.pdf> (last accessed 6 April 2013).

81 Available at <http://unfccc.int/resource/docs/napa/eth01.pdf> (last accessed 6 April 2013).

the malaria containment programme; and promoting farm and homestead forestry and agroforestry practices in arid, semiarid and dry sub-humid parts of Ethiopia. The NAPA was updated and replaced in 2010 by the Ethiopian Programme of Adaptation to Climate Change, which is tasked to identify risks related to climate change and oversee institutions in charge of mitigating these risks.

4.8.3 Kenya[82]

The 2010 Kenyan Constitution[83] provides a modern framework for environmental rights. In its preamble it states that the people of Kenya are respectful of the environment, "which is our heritage, and determined to sustain it for the benefit of future generations." In the section on rights and fundamental freedoms, the Constitution in Article 42 recognizes that everyone has a "right to a clean and healthy environment," which includes the right

(a) to have the environment protected for the benefit of present and future generations through legislative and other measures, particularly those contemplated in Article 69; and

(b) to have obligations relating to the environment fulfilled under Article 70.

Under Chapter 5 (Land and Environment) of the Constitution, the enforcement of environmental rights is regulated in Article 70(1), where it states that if a person alleges that a right to a clean and healthy environment recognized and protected under Article 42 has been, is being or is likely to be, denied, violated, infringed or threatened, the person may apply to a court for redress in addition to any other legal remedies that are available in respect to the same matter.

Kenya signed the UNFCCC in 1992, ratifying it in 1994, along with the Kyoto Protocol in 2005. Kenya (a nonAnnex I country) has placed great importance on issues of climate change, for example its 2008 Draft National Environment Policy, which recognizes that climate change involves many ministries, gives climate change a higher profile at a national level to help address climate-related issues in respect of the relevant ministries.[84]

Kenya has developed a Climate Change Authority Act that is making progress through parliament and is likely to be voted into law in early 2013. In November 2012 a public validation process paved the way for government to approve a land launch in early 2013 of a complementary Climate Change Action Plan that defines clear measures on adaptation and mitigation, including nationally appropriate mitigation actions, a low-carbon development

82 Based largely on GLOBE International 2013, pp. 269-277.

83 *See* C. Glinz, 'Kenya's New Constitution: A Transforming Document or Less Than Meets the Eye?', *Verfassung und Recht in Übersee*, Vol. 44, 2011, pp. 60-80.

84 *See* P. Kameri-Mbote & O. Collins, 'Kenya', *in* R. Lord *et al.* (Eds.), *Climate Change Liability. Transnational Law and Practice*, Cambridge, Cambridge University Press, 2012, pp. 296-318.

strategy, knowledge management and capacity development, financing mechanisms and the creation of an institutional structure to ensure effective coordination. Developed by the Kenyan Government, through the Ministry of Environment and Mineral Resources, and in conjunction with donor partners, the Action Plan provides a platform for the implementation of the 2010 National Climate Change Response Strategy. When it is passed, the Action Plan could qualify to be considered Kenya's flagship legislation on climate change. Another process under way is the formulation of the National Environment Policy, also expected to be adopted soon. The Climate Change Authority Bill was introduced in the National Assembly in 2012. In addition to establishing a Climate Change authority, the Bill sets out to provide for the development of strategies to address the effects of climate change, as well as to forge a framework for mitigation of and adaptation to climate.

4.8.4 Namibia[85]

The 1990 Namibian Constitution, many international treaties, as well as a multitude of statutory enactments and policies provide for a wide field of environmental protection in Namibia. Over the past years, a bundle of new environmental legislation has been passed, and it thus becomes evident that environmental concern has gained momentum. The Namibian environmental policy framework[86] determines the guiding objectives and the strategies to be used to strengthen the respect for environmental values, taking into account the existing social, cultural and economic situation. The foundation for the Namibian environmental policy framework is Article 95(l) of the Namibian Constitution of 1990. It stipulates that the state shall actively promote and maintain the welfare of the people by adopting policies that include the "maintenance of ecosystems, essential ecological processes and biological diversity of Namibia and utilization of living natural resources on a sustainable basis [. . .]."

It is worthwhile mentioning that the Namibian Constitution explicitly incorporates international law and makes it part of the law of the land. *Ab initio*, public international law is part of the law of Namibia. No transformation or subsequent legislative act is needed. A treaty will become binding upon Namibia in terms of Article 144 of the Constitution if the relevant international and constitutional requirements have been met.[87]

Namibia ratified the UNFCCC in 1995 and became legally obliged to adopt and implement policies and measures designed to mitigate the effects of climate change and to adapt to such changes. Namibia acceded to the Kyoto Protocol in 2003. Namibia's Initial National

85 Based largely on Ruppel & Ruppel-Schlichting, 2013.
86 *Ibid.*
87 *See* with further references Ruppel, 2010a.

Communication to the Conference of Parties of the UNFCCC was submitted in 2002 in accordance with decisions taken at various COPs to the UNFCCC. The Ministry of Environment and Tourism (MET) through the Directorate of Environmental Affairs (DEA) is responsible for overseeing the coordination of climate change issues in Namibia.[88]
Vision 2030, the various National Development Plans, various sector policies and Cabinet directives are in place. Moreover, Namibia in 2008 drew up a draft national climate change strategy and action plan – which was introduced in 2009.[89] In June 2011 Namibia's Prime Minister, Nahas Angula, indicated that a national policy on climate change for Namibia should be placed within the global framework of political, developmental and technological interests.[90] The then Namibian minister of Environment and Tourism made the following significant statement: "While climate change has the potential to side-rail development processes, the key is to prepare sufficiently and effectively and to use the threats and opportunities of climate change to lay the basis for sustainability and prosperity."[91]

4.8.5 Rwanda[92]

Article 49 (I) of the 2003 Rwandan Constitution stipulates that every citizen is entitled to a healthy and satisfying environment. This provision obliges the government to commit itself to ensuring that citizens live in a safe and clean environment. To achieve this obligation, reforestation and terracing are emphasized on hilly areas to prevent soil erosion, while marshlands and low-land vegetation are protected. Article 49(II) provides that every person has a duty to protect, safeguard and promote the environment, and places the State under the obligation to protect the environment. Article 49(III) stipulates that the law shall determine the modalities for protecting, safeguarding and promoting the environment.
Rwanda has ratified almost all international instruments related to the protection of the environment. The government provides orientation through the national policy on environment and the national policy on water and sanitation. The Rwanda Environment Management Authority takes the lead in regulating, safeguarding and promoting safe and clean environment protection programmes.
Rwanda ratified the UNFCCC and the Kyoto Protocol in 1998. In 2005, Rwanda submitted its initial report to the UNFCCC, and in June 2012 its second communication, including

88 I. Mapaure, 'Climate Change in Namibia: Projected Trends and Effects,' *in* Ruppel & Ruppel-Schlichting, 2013.

89 *See* J.K. Mfune *et al.*, *Namibia Climate Change Strategy and Action Plan*. Windhoek, Ministry of Environment and Tourism, 2009, available at <www.met.gov.na/Documents/NAMIBIA-proposed%20Climate%20Change%20Strategy%20and%20Action%20Plan%20(13.pdf> (last accessed 13 January 2013).

90 *See* <www.namibian.com.na/news-articles/national/full-story/archive/2011/june/article/policy-on-climate-change-needs-review-pm/> (last accessed 18 February 2013).

91 N. Nandi-Ndaitwah, 'The Good and Bad of Climate Change', *The Namibian*, 16 June 2011.

92 Based largely on GLOBE International, 2013, pp. 365-371.

a stand-alone mitigation strategy, the Carbon Policy and an updated emissions inventory. The Constitution of 2003 is Rwanda's supreme law. In 2011 the Government published the National Climate Change and Low-Carbon Development Strategy. The Action Plan for the Ministry of Natural Resources July 2011-June 2012 sets specific targets for reducing climate change vulnerability. In May 2012, a law establishing a national fund for climate change financing, FONERWA, was passed and is expected to contribute to Rwanda's existing financing gap. In 2009 a Climate Change and International Obligations Unit was established to coordinate carbon market activities. Rwanda's climate change efforts are supported by various international donors. In 2010, the Government of Rwanda, the Government of Japan, the United Nations Development Programme and the United Nations Environment Programme (UNEP) launched two climate change adaptation programmes. One focuses on reducing vulnerability to climate change by establishing early warning and disaster preparedness systems and support for integrated watershed management in flood-prone areas; the other focuses on building an integrated comprehensive national adaptation approach in Rwanda.

4.8.6 *South Africa*[93]

Section 24 of the 1996 South African Constitution states, among others, that everyone has the right

(a) to an environment that is not harmful to their health or well-being; and
(b) to have the environment protected, for the benefit of present and future generations, through reasonable legislative and other measures that
    (i) prevent pollution and ecological degradation;
    (ii) promote conservation; and
    (iii) secure ecologically sustainable development and use of natural resources while promoting justifiable economic and social development.

The Constitution also provides the framework for the administration of environmental laws. It designates the environment as a matter of concurrent national and provincial responsibility. Environmental law is contained in a multitude of statutes and regulations.[94] South Africa signed the UNFCCC in 1993 and ratified it in 1997. It acceded to the Kyoto Protocol in July 2002. Within the Sub-Directorate for Global Climate Change of the

93 Based on GLOBE International, 2013, pp. 373-379.

94 Relevant to the context of climate change are, *inter alia,* the following statutes: National Environmental Management Act, 1998 (Act 107 of 1998); National Water Act, 1998 (Act 36 of 1998); Marine Living Resources Act, 1998 (Act 18 of 1998); Disaster Management Act, 2002 (Act 57 of 2002); National Environmental Management: Biodiversity Act, 2004 (Act 10 of 2004); National Environmental Management: Protected Areas Act, 2003 (Act 57 of 2003); National Environmental Management: Waste Management Act 2008; and National Environmental Management: Air Quality Act, 2004 (Act 39 of 2004).

Department of Environmental Affairs lies the responsibility to ensure that South Africa complies with its obligations under the UNFCCC. In 2011, South Africa's Second National Communication under the UNFCCC was published.[95]

South Africa has dealt with climate change largely through policies, strategies and regulations. Through these measures, South Africa is showing its commitment to tackling climate change, particularly in developing market-based mitigation mechanisms and promoting renewable energy and energy efficiency.

The 2004 National Climate Change Response Strategy represented the first direct recognition of the need for action on climate change. Two years later, the Cabinet commissioned the Long-Term Mitigation Scenario study in an attempt to produce sound scientific analysis from which the Government could derive a long-term climate policy. The study produced a series of policy recommendations. In July 2008, the Vision, Strategic Direction and Framework for Climate Policy were announced.

The current flagship policy in South Africa is the National Climate Change Response Policy, approved by Cabinet in October 2011. This policy's White Paper presents the South African Government's vision for an effective climate change response and the long-term, just transition to a climate resilient and lower carbon economy and society. It reflects a strategic approach referred to as "climate change resilient development," addressing both adaptation and mitigation. The White Paper accepts the conclusions of the Intergovernmental Panel on Climate Change (IPCC), regards climate change as one of the greatest threats to sustainable development, reaffirms its commitment towards the UNFCCC and the Kyoto Protocol and undertakes to develop a comprehensive national response plan, of which the White Paper is an integral part.

The Taxation Law Amendment Bill of 2009 amends the 1962 Income Tax Act to include, among other things, income tax incentives for participation in Clean Development Mechanism projects as well as for energy efficiency savings. The Clean Development Mechanism projects are run by a designated national authority established under the Department of Energy, and they are governed by regulations published under the National Environmental Management Act of 1998. During the 2012-2013 budget discussions in February 2012, the minister of Finance introduced a proposed carbon tax on annual emissions for all sectors, including electricity, petroleum, iron, steel and aluminium.

4.9 DEVELOPMENTS AND GAPS

The aforementioned passages reflect that much is already being done in Africa: the rolling out of AU policy pathways and related legal processes, the accommodation of national

95 *See* J. Glazewski & D. Collier, 'South Africa', *in* R. Lord *et al.* (Eds.), *Climate Change Liability. Transnational Law and Practice*, Cambridge, Cambridge University Press, 2012, pp. 319-348.

constitutional and legislative frameworks and the development of climate change policy are all in progress and are also critical for the continent. Resolution (A/RES/67/1) was recently adopted by the UN General Assembly on the Rule of Law – which underlines the importance of fair, stable and predictable legal frameworks for generating inclusive, sustainable and equitable development and for maintaining peace and security. In this light, it is also essential to further the discussions on

- the linkage between climate change, sustainable development and the rule of law
- the mutually supportive relationship between the adherence to the rule of law and the respect for all human rights, including those related to the environment
- access to justice in environmental matters and new and emerging principles, such as the nonregression in environmental law and
- the importance of country systems and other developments at the national level to improve the effectiveness of environmental and climate change law.

Many African governments have progressed in addressing climate change and related issues. The AU has also succeeded in presenting an increasingly cohesive African position on climate change. However, significant gaps remain.[96]

At the regional level, effectively managing consensus and divergence remains challenging. Divergent priorities among African countries threaten the potential of the AU to wield influence in international climate politics. The AU thus needs to address divergence factors in lieu of conducting international negotiations.

At the national and sub-regional levels, various policy gaps exist, a major one being the sluggishness in bringing climate change into the mainstream planning of all development sectors. In many African countries, a climate policy is non-existent or still in the making. Development and climate policy run parallel, and integrated development-climate policy framework is non-existent, making it difficult to stop the rebranding of Official Development Assistance (ODA) as climate response. Very few sectoral policies consider climate change and need reviewing to close this gap.[97]

Moreover, African governments should make provisions for integrating climate change responses into national budgetary allocations. Economic planners often lack guidelines on mainstreaming climate change adaptation at the national level. As with other policy spheres, climate-proofing development through integrating climate change in all policy spheres has its costs and trade-offs. Considering the chronic understaffing and

96 This passage is based on O.C. Ruppel & C.I. Speranza, 'The Institutional, Legal and Policy Frameworks, *in* AMCEN/African Ministerial Conference on the Environment, *Guidebook: Addressing Climate Change Challenges in Africa*, Nairobi: African Union, AMCEN & United Nations Environment Programme, pp. 138-166, available at <www.unep.org/roa/amcen/docs/publications/guidebook_CLimateChange.pdf> (last accessed 23 January 2013).

97 *Ibid.*

underfunding of certain government activities, climate change will bring more work, and governments should improve staff skills and provide more resources to address the add-on challenges of climate change. Many operational limitations hinder implementation. These include dysfunctional organizational arrangements causing conflicting and overlapping mandates, overburdening reporting requirements of various international agreements and conventions, and inability to retain skilled staff.

Lastly, designing policies for dealing with climate change offers an opportunity to address the dichotomy between parallel regulatory systems, *i.e.* the customary and the state laws, *e.g.* in access to land, the aspects of the management, use of and control over natural resources and benefit sharing. However, issues remain on how to develop a national framework for compensating natural resource users for providing environmental services and how to proceed with a low-carbon development, renewable energy schemes and the role of green transformation in these processes.[98]

4.10 CONCLUSION

Although it is primarily the responsibility of developed countries to reduce their GHG emissions first, in line with the principle of common but differentiated responsibility, developing countries also need to make specific policy commitments. What became very clear during the past few years is that we live in a world that is very different from the context in which the Kyoto Protocol was established. Compared with the mid-1990s, the debates are now taking place in a much changed world. This applies to Africa, but even more notably to China and other developing countries that have gained more political clout and greater impact on the world's climate. More effective action is still needed.[99]

For Africa, climate change is becoming an increasingly threatening reality, but one that is often too narrowly perceived in economic and sectoral policies, meaning that the severe negative consequences on the people are still largely being neglected.[100]

According to IPCC findings[101]

> Africa is one of the most vulnerable continents to climate change and climate variability, a situation aggravated by the interaction of "multiple stresses," occurring at various levels, and low adaptive capacity [. . .]. Africa's major economic sectors are vulnerable to current climate sensitivity, with huge

98 *Ibid.*

99 *See* Ruppel *et al.*, 2011.

100 P.M. Susswein, 'The Use of Market Based Economic Instruments for Sustainable Environmental Management in Africa: A Case Study of the SADC Region, *in* B. Chaytor & K.R. Gray (Eds.), *International Environmental Law and Policy in Africa,* Dordrecht, Kluwer, 2003, p. 297.

101 Boko *et al.*, 2007, p. 435.

> economic impacts, and this vulnerability is exacerbated by existing developmental challenges such as endemic poverty, complex governance and institutional dimensions; limited access to capital, including markets, infrastructure and technology; ecosystem degradation; and complex disasters and conflicts. These in turn have contributed to Africa's weak adaptive capacity, increasing the continent's vulnerability to projected climate change.

In response thereto, the AU has introduced initiatives aiming to ensure that poor and marginalized communities in Africa, *i.e.* women, children and indigenous peoples, do not suffer a disproportionate burden associated with climate change. The same applies to various African countries, where significant (legal) developments related to climate change policy reflect political will towards addressing climate change.[102]

The AU has been decisive on action. Populations whose rights are poorly protected are likely to be less well equipped to understand or prepare for climate change; they would be less able to enforce their rights or lobby them effectively and are thus more likely to be unable to adapt to the anticipated changes in their environmental and economic situation. What remains essential is that Africa is very clear on the fact that the majority of its people are vulnerable, *i.e.* subjected to displacement, landlessness, joblessness, homelessness, marginalization, food insecurity, increased morbidity, loss of access to common property resources and social disarticulation. These people are particularly vulnerable to extreme events, on the basis of a wide range of social vulnerability characteristics.[103] This needs to be brought more sharply to the attention of the global community in order to develop more international solidarity in terms of a global corporate social responsibility to protect those populations that need it most. In the same light, it is hoped that in future we can come up with more global and consolidated approaches when it comes to water, climate, energy, economy, security and development policy. In the interest of Africa and its people, this – without doubt – requires still more international cooperation, regional integration and local implementation.[104] In fact, a new global "social contract refers to the necessity of humankind taking collective responsibility for the avoidance of dangerous climate change [. . .] to the planet."[105]

Moreover, addressing climate change adequately also requires substantial investment in new technologies, processes and services. Therefore, a favourable investment climate is

102 Ruppel & Speranza, 2011, pp. 199-202.

103 Intergovernmental Panel on Climate Change, 'Managing the Risks of Extreme Events and Disasters to Advance Climate Change Adaptation. A Special Report of Working Groups I and II of the Intergovernmental Panel on Climate Change', *in* C.B. Field *et al.* (Eds.), Cambridge, Cambridge University Press, 2012.

104 *See* Ruppel, 2012b for further references.

105 Wissenschaftlicher Beirat der Bundesregierung Globale Umweltveränderungen Flagship Report, *World in Transition: A Social Contract for Sustainability, German Advisory Council on Global Change*, Berlin,WBGU, 2011, p. 8.

pivotal, and adequate framework conditions for more inclusive climate investment, leveraging private sector resources and seizing opportunities for innovation will be needed. In light of this, Africa now needs to strengthen its long-term vision of working together on the African continent, with global partners; and, in adapting to climate change, it needs to promote climate investment, sustainable economic development and green economic growth.

A major problem unfortunately prevails. The World Corruption Report 2011[106] examined this issue and stated that corruption was indeed a risk in addressing climate change, since a risk of corruption always exists where "huge amounts of money flow through new and untested financial markets and mechanisms". The reasons for the high risk of corruption with regard to climate finance are rooted in the level of complexity, uncertainty and novelty that surrounds many climate issues. A multitude of regulatory grey zones and loopholes exist that are at risk of being exploited to satisfy corrupt interests. To ensure that the investments by the public and private sectors are properly and equitably managed, a system of good climate governance[107] with participatory, accountable, transparent, inclusive and responsive policy development and decisions, and respect for the rule of law, is essential.

106 Transparency International, *World Corruption Report: Climate Change*, London, Earthscan, 2001.

107 'Climate governance' can be understood as the processes that currently exist at the international, national, corporate and local levels to address the causes and effects of climate change. *See* Transparency International, 2011, p. 3.

Bibliography

AMCEN/African Ministerial Conference on the Environment, *Addressing Climate Change Challenges in Africa: A Practical Guide towards Sustainable Development*, 2011. Retrieved on 15 January 2013 from <www.unep.org/roa/amcen/docs/.../guidebook_CLimateChange.pdf>.

Boko, M. *et al.*, 'Africa', *in* Parry, M.L. *et al.* (Eds.), *Climate Change 2007: Impacts, Adaptation and Vulnerability. Contribution of Working Group II to the Fourth Assessment Report of the Intergovernmental Panel on Climate Change*, Cambridge, Cambridge University Press, 2007, pp. 433-467.

Eze, O.C., 'The Conventions Governing the Specific Aspects of Refugee Problems in Africa, *in* Yusuf, A.A. & Fatash. O. (Eds.), *The African Union: Legal and Institutional Framework*, Leiden; Boston, Martinus Nijhoff Publishers, 2012, pp. 495-518.

FAO/Food and Agricultural Organization of the United Nations, *State of the World's Forests 2011*, Rome, FAO, 2011.

Federal Democratic Republic of Ethiopia, *Ethiopia's Climate-Resilient Green Economy Strategy*, Addis Ababa: Environmental Protection Agency, 2011. Retrieved on 6 April 2013 from <www.uncsd2012.org/content/documents/287CRGE%20Ethiopia%20Green%20 Economy_Brochure.pdf>.

Glazewski, J. & Collier, D., 'South Africa', *in* Lord, R. *et al.* (Eds.), *Climate Change Liability. Transnational Law and Practice*, Cambridge, Cambridge University Press, 2012, pp. 319-348.

Glinz, C., 'Kenya's New Constitution: A Transforming Document or Less Than Meets the Eye?', 44 *Verfassung und Recht in Übersee*, 2011, pp. 60-80.

GLOBE International, *Climate Legislation Study: A Review of Climate Change Legislation in 33 Countries* (3rd edn), *in* Townshend, T. *et al.* (Eds.), 2013. Retrieved on 18 February 2013 from <www.globeinternational.org/images/climate-study/3rd_GLOBE_Report.pdf>.

Government of the Democratic Republic of Congo, *Programme d'Action National d'Adaptation au Changement Climatique de la République Démocratique du Congo*, 2006. Retrieved on 18 January 2012 from <http://unfccc.int/resource/docs/napa/cod01.pdf>.

Gray, K.R. (Eds.), *International Environmental Law and Policy in Africa*, Dordrecht, Kluwer, 2003.

Hoste, J-C., *Where Was United Africa in the Climate Change Negotiations? Africa Policy Brief No. 2*, Brussels: Egmont Royal Institute for International Relations, 2010. Retrieved on 13 January 2013 from <www.edc2020.eu/fileadmin/Textdateien/post_COP_15_briefing/Jean_Christophe_Hoste_-_Where_was_united_Africa_in_the_climate_change_negotiations_-_EDC_2020.pdf>.

IPCC/Intergovernmental Panel on Climate Change, 'Managing the Risks of Extreme Events and Disasters to Advance Climate Change Adaptation. A Special Report of Working Groups I and II of the Intergovernmental Panel on Climate Change', *in* Field, C.B. *et al.* (Eds.), Cambridge, Cambridge University Press, 2012.

IRENA/International Renewable Energy Agency, *Prospects for the African Power Sector*, 2011. Retrieved on 17 January 2013 from <www.irena.org/DocumentDownloads/Publications/Prospects_for_the_African_Power_Sector_abstract.pdf>.

Kabau, T., 'The Responsibility to Protect and the Role of Regional Organizations: An Appraisal of the African Union's Interventions', 4 *Goettingen Journal of International Law*, 2012, pp. 49-92.

Kameri-Mbote, P. & Odote, C., 'Kenya', *in* Lord, R. *et al.* (Eds.), *Climate Change Liability. Transnational Law and Practice*, Cambridge, Cambridge University Press, 2012, pp. 296-318.

Keohane, R.O. & Victor, D.G., 'The Regime Complex for Climate Change', 9(1) *Perspective on Politics*, 2011, pp. 7-23.

Koroma, A.D., 'Solidarity: Evidence of an Emerging International Legal Principle', *in* Hestermeyer, H.P., *et al.* (Eds.), *Coexistence, Cooperation and Solidarity: Liber Amicorum Rüdiger Wolfrum*, Leiden, Martinus Nijhoff Publishers, 2012, pp. 103-129.

Mapaure, I., 'Climate Change in Namibia: Projected Trends and Effects', *in* Ruppel, O.C. & Ruppel-Schlichting, K. (Eds.), *Environmental Law and Policy in Namibia* (2nd edn), Windhoek, Orumbonde Press; Essen, Welwitschia Verlag Dr. A. Eckl, 2013, pp. 245-264.

Mfune, J.K. *et al.*, *Namibia Climate Change Strategy and Action Plan*, Windhoek, Ministry of Environment and Tourism, 2009. Retrieved on 13 January 2013 from <www.met.gov.na/Documents/NAMIBIA-proposed%20Climate%20Change%20Strategy%20and%20Action%20Plan%20(13.pdf>.

Ruppel, O.C. (Ed.), *Women and Custom in Namibia: Cultural Practice versus Gender Equality?*, Windhoek, Macmillan Education Namibia, 2008. Retrieved on 13 January 2012 from <www.kas.de/proj/home/pub/8/2/dokument_id-15086/index.html>.

Ruppel, O.C., 'Regional Economic Communities and Human Rights in East and Southern Africa', *in* Bösl, A. & Diescho, J. (Eds.), *Human Rights in Africa* Windhoek, Macmillan Publishers Namibia, 2009a, pp. 273-347.

Ruppel, O.C., 'The Protection of Children's Rights under International Law from a Namibian Perspective', *in* Ruppel, O.C. (Ed.), *Children's Rights in Namibia*, Windhoek, Macmillan Publishers Namibia, 2009b, pp. 53-100.

Ruppel, O.C., 'Environmental Rights and Justice in Namibia', *in* Bösl, A., Horn, N. & du Pisani, A. (Eds.), *Constitutional Democracy in Namibia: A Critical Analysis after Two Decades*, Windhoek, Macmillan Education Namibia, 2010a, pp. 323-360.

Ruppel, O.C., *Women's Rights and Customary Law in Namibia: A Conflict between Human and Cultural Rights?*, Basler Afrika Bibliographien BAB Working Paper, 2010b. Retrieved on 20 January 2012 from <www.baslerafrika.ch/upload/files/WP_2010_2_Ruppel.pdf>.

Ruppel, O.C., 'International Environmental Law from a Namibian Perspective', *in* Ruppel, O.C., & Ruppel-Schlichting, K. (Eds.), *Environmental Law and Policy in Namibia*, Windhoek, Orumbonde Press, 2011a, pp. 33-78.

Ruppel, O.C., 'Human Rights and the Environment', *in* Ruppel, O.C. & Ruppel-Schlichting, K. (Eds.), *Environmental Law and Policy in Namibia*, Windhoek, Orumbonde Press, 2011b, pp. 219-237.

Ruppel, O.C., 'Climate Change Policy Positions and Related Developments in the AU and SADC', 2(1) *SADC Law Journal*, 2012a, pp. 14-35.

Ruppel, O.C., 'Wasser und Land. Brennpunkte innerhalb der Entwicklungsgemeinschaft des südlichen Afrika (SADC)', *in* Reder, M. & Pfeifer, H. (Eds.), *Kampf um Ressourcen. Weltordnung zwischen Konkurrenz und Kooperation (Globale Solidarität – Schritte zu Einer Neuen Weltkultur, 22*, Stuttgart, Kohlhammer, 2012b, pp. 59-85.

Ruppel, O.C., Grimm, S. & van Wyk, S., *COP 17: Addressing Climate Change in a Changed World – Chinese and African Positions*, 2011. Retrieved on 7 February 2013 from <www.ccs.org.za/wp-content/uploads/2011/12/COP17.pdf>.

Ruppel, O.C. & Speranza, C.I., 'The Institutional, Legal and Policy Frameworks', *in* AMCEN/African Ministerial Conference on the Environment, *Guidebook: Addressing Climate Change Challenges in Africa*, Nairobi, African Union, AMCEN & United Nations Environment Programme. 2011. Retrieved on 23 January 2013 from <www.unep.org/roa/amcen/docs/publications/guidebook_CLimateChange.pdf>, pp. 138-166.

Ruppel, O.C. & van Wyk, S., 'The Effects of Climate Change on Human Security and Human Rights in Africa', *China Monitor*, 2011. Retrieved on 23 December 2012 from <www.ccs.org.za/wp-content/uploads/2011/10/China_Monitor_OCT_2011_final.pdf>.

Ruppel, O.C. & Ruppel-Schlichting, K., 'Climate Change and Human Security: Relevant for Regional Integration in SADC?', *in* Hartzenberg, T., Erasmus, G. & du Pisani, A. (Eds.), *Monitoring Regional Integration in Southern Africa, 2011 Yearbook for Regional Integration*, Stellenbosch, Trade Law Centre for Southern Africa (TRALAC), 2012, pp. 32-71.

Ruppel, O.C. & Bwiza, D., 'Country Report: Democratic Republic of Congo: Recent Developments in Environmental Protection', 1 *IUCN E-Journal*, 2013 (forthcoming).

Ruppel, O.C. & Ruppel-Schlichting, K. (Eds.), *Environmental Law and Policy in Namibia* (2nd edn), Windhoek, Orumbonde Press; Essen, Welwitschia Verlag Dr. A. Eckl, 2013.

Scholtz, W., 'The Promotion of Regional Environmental Security and Africa's Common Position on Climate Change', 10 *African Human Rights Law Journal*, 2010, pp. 1-25.

Scott, S.V., 'The Securitization of Climate Change in World Politics: How Close Have We Come and Would Full Securitization Enhance the Efficacy of Global Climate Change Policy?', 21(3) *Review of European Community and International Environmental Law*, 2012, pp. 220-230.

Susswein, P.M., 'The Use of Market Based Economic Instruments for Sustainable Environmental Management in Africa: A Case Study of the SADC Region', *in* Chaytor, B. & Gray, K.R. (Eds.), *International Environmental Law and Policy in Africa*, Dordrecht, Kluwer, 2003.

Thielbörger, P. & Liburd, M., *Haiti Two Years after the Earthquake – A Case for the Responsibility to Protect?*, Institute for International Law of Peace and Armed Conflict, BOFAXE No. 399E, 2012.

Transparency International, *World Corruption Report: Climate Change*, London, Earthscan, 2011.

UN/United Nations Womenwatch. *Fact Sheet: Women and Climate Change*, 2009. Retrieved on 25 June 2011 from <www.un.org/womenwatch/feature/climate_change/downloads/Women_and_Climate_Change_Factsheet.pdf>.

UNEP/United Nations Environmental Programme, *Africa Atlas of Our Changing Environment*, Nairobi, UNEP, 2008. Retrieved on 28 December 2012 from <http://unep.org/dewa/africa/africaAtlas/>.

UNFCCC/United Nations Framework Convention on Climate Change, *Report of the Green Climate Fund to the Conference of the Parties and Guidance to the Green Climate Fund*. 2012. Retrieved on 18 February 2013 from <http://unfccc.int/resource/docs/2012/cop18/eng/05.pdf>.

UNICEF/United Nations Children's Fund, *A Brighter Tomorrow: Climate Change, Child Rights and Intergenerational Justice*, 2009. Retrieved on 25 June 2011 from <www.unicef.org.uk/Documents/Publications/intergenerationaljustice.pdf>.

Van der Linde, M. & Louw, L., 'Considering the Interpretation and Implementation of Article 24 of the African Charter on Human and Peoples' Rights in Light of the SERAC Communication', 3(1) *African Human Rights Law Journal*, 2003, pp. 167-187.

WBGU/Wissenschaftlicher Beirat der Bundesregierung Globale Umweltveränderungen, *Flagship Report, World in Transition: A Social Contract for Sustainability, German Advisory Council on Global Change*, Berlin, WBGU, 2011.

Editorial board

9 789462 360808

13W71306/ T1/ 9789462360808